Dynamics of Freezing-Thawing Soil around Subway Shield Tunnels

Zhen-Dong Cui · Zhong-Liang Zhang ·
Zhi-Xiang Zhan · Peng-Peng He ·
Chen-Yu Hou · Yan-Kun Zhang

Dynamics of Freezing-Thawing Soil around Subway Shield Tunnels

Springer

Zhen-Dong Cui
State Key Laboratory for Geomechanics and Deep Underground Engineering, School of Mechanics and Civil Engineering
China University of Mining and Technology
Xuzhou, China

Zhong-Liang Zhang
State Key Laboratory for Geomechanics and Deep Underground Engineering, School of Mechanics and Civil Engineering
China University of Mining and Technology
Xuzhou, China

Zhi-Xiang Zhan
State Key Laboratory for Geomechanics and Deep Underground Engineering, School of Mechanics and Civil Engineering
China University of Mining and Technology
Xuzhou, China

Peng-Peng He
State Key Laboratory for Geomechanics and Deep Underground Engineering, School of Mechanics and Civil Engineering
China University of Mining and Technology
Xuzhou, China

Chen-Yu Hou
State Key Laboratory for Geomechanics and Deep Underground Engineering, School of Mechanics and Civil Engineering
China University of Mining and Technology
Xuzhou, China

Yan-Kun Zhang
State Key Laboratory for Geomechanics and Deep Underground Engineering, School of Mechanics and Civil Engineering
China University of Mining and Technology
Xuzhou, China

Funded by the Fundamental Research Funds for the Central Universities (Grant No. 2018ZZCX04) and the National Key Research and Development Program (Grant No. 2017YFC1500702).

ISBN 978-981-15-4344-9 ISBN 978-981-15-4342-5 (eBook)
https://doi.org/10.1007/978-981-15-4342-5

This Springer imprint is published by the registered company Springer Nature Singapore Pte Ltd.
The registered company address is: 152 Beach Road, #21-01/04 Gateway East, Singapore 189721, Singapore

Preface

During the operation of the subway, the vibration caused by the subway cyclic loading exists for a long time, which will reduce the strength of the soft soil foundation, generate different degrees of settlement and deformation, affect the deformation of the tunnel axis, and cause ground subsidence. The traffic facilities of many subway construction projects are usually built on soft soil foundation. In the initial stage of operation after the construction of freezing method, due to the action of cyclic loading, consolidation deformation settlement often occurs, which has great safety problems.

Supported by the Fundamental Research Funds for the Central Universities (Grant No. 2018ZZCX04) and the National Key Research and Development Program (Grant No. 2017YFC1500702), this monograph studies the dynamic characteristics of soil around subway shield tunnel by artificial freezing method, which can provide theoretical basis and guidance for the effective control of long-term subway settlement and maintaining the safety of subway operation.

Chapter 1 is Introduction. It introduces the progresses of the mechanical properties of soft clay under cyclic loading, the physical and mechanical properties of soft clay under freeze-thaw cycle and the macroscopic and microstructural characteristics of soft clay.

Chapter 2 is Artificial freezing for by-pass of subway tunnel. In this chapter, the artificial freezing method used in the construction of the by-pass in the silty clay stratum is introduced. The design of freezing parameters, construction technologies, monitoring of the temperature field and deformation are introduced systematically.

Chapter 3 is Axial strain of silty clay before and after freezing and thawing. In this chapter, a series of cyclic triaxial tests have been carried out to investigate the dynamic behavior of soft clays before and after freezing-thawing. The stress-strain backbone loop, axial strain and excess pore pressure of the grey muck clay are evaluated and compared under different frequencies and cyclic stress ratios.

Chapter 4 is Freezing-thawing on dynamic characteristics of silty clay. In this chapter, 12 cyclic triaxial tests have been carried out to investigate the dynamic characteristics of the silty clay before and after freezing-thawing, including the dynamic modulus, the damping ratio, the dynamic strain and the excess pore water pressure, etc.

Chapter 5 is Microscopic pore structures of silty clay before and after freezing-thawing by SEM. The parameters of microscopic pore structure of the silty clay, including the pore shape coefficient, the pore orientation anisotropy, the pore size distribution and the fractal dimension, are evaluated with different frequencies, cyclic stress ratios and freezing temperatures.

Chapter 6 is Microscopic pore structures of silty clay before and after freezing-thawing by MIP. In this chapter, 14 MIP tests following the cyclic triaxial tests of the silty clay of layer No. 5 in Shanghai were conducted to investigate the variation of the PSD of the silty clay. The effects of freezing-thawing and cyclic loadings on the PSD of the silty clay were evaluated and the MIP derived water retention curves of the silty clay were predicted with the van Genuchten model.

Chapter 7 is Dynamics of silty clay around subway shield tunnel before and after freezing and thawing by numerical simulation. In this chapter, the three-dimensional governing equations were derived and the solving method was given. The material parameters of the thawing soil and the undisturbed soil used in the numerical simulations were obtained by dynamic triaxial tests. The dynamic response of thawing soil and undisturbed soil in the same environmental conditions were compared and analyzed.

Chapter 8 is Conclusions and Prospects. This part comprehensively summarizes the research conclusions. Several controversial issues are discussed and then the further research work and prospects are simply described.

This monograph has been prepared with the combined effort of all researchers in the group under Prof. Zhen-Dong Cui's leading, in which Zhong-Liang Zhang, Zhi-Xiang Zhan, Chen-Yu Hou, Yan-Kun Zhang and some other students all have involved in this comprehensive research work in this monograph.

Xuzhou, People's Republic of China Prof. Zhen-Dong Cui
February 2020

Contents

About the Author

Prof. Dr. Zhen-Dong Cui is the Dean of Institute of Geotechnical Engineering, China University of Mining and Technology. In July 2008, he obtained Ph.D. from School of Civil Engineering, Tongji University, Shanghai, China. Since then, he had been a postdoctoral research fellow in the Hong Kong University of Science and Technology for one year. He joined Shanghai Institute of Geological Survey in 2009. In July 2010, he joined CUMT as an associate professor and was promoted to full professor in 2013. Supported by China Scholarship Council, from Aug. 2015 to Aug. 2016, as a visiting scholar, he researched and studied in the Department of Civil, Environmental and Architectural Engineering, University of Colorado Boulder.

He won Nomination of 100 Excellent Doctoral Dissertations in China in 2011. He won Shanghai Excellent Doctoral Dissertations in 2010 and Excellent doctorate thesis of Tongji University in 2009. In 2015, he was awarded the third prize of Shanghai Natural Science. In 2013, he was awarded the second prize of Natural Science of the Ministry of Education. In 2008, he was awarded the second prize of Progress of Science and Technology in Shanghai. In 2012, he was selected as Qinglan Project for Outstanding Young Teachers of Jiangsu Province and in 2016, he was selected as 333 Talent Project in Jiangsu Province. In 2014 and in 2017, he was twice selected as Young Academic Leader of

China University of Mining and Technology. In 2015, he was awarded as Excellent Innovation Team Leader of China University of Mining and Technology.

His research interests focus mainly on the urban subway tunnel, the deformation of soft foundation, the soil dynamics, the centrifuge model test and the artificial frozen soil. In the past 5 years, he took charge of National Natural Science Foundation of China (NSFC), Jiangsu Natural Science Foundation of China, Outstanding Innovation Team Project in China University of Mining and Technology and Special Fund for China Postdoctoral Science Foundation. He published more than 60 papers, in which 38 English papers (indexed by SCI) have been published in Engineering Geology, Natural Hazards, Computers and Geotechnics, Cold Regions Science and Technology, Environmental Earth Sciences, International Journal of Rock Mechanics and Mining Sciences, etc. He has applied for 10 national invented patents, among which five patents have been awarded. He is the specialized committee member of the soft soil engineering of the geotechnical engineering branch of the Civil Engineering Society of China, the member of International Society for Soil Mechanics and Geotechnical Engineering (ISSMGE) and the member of International Association for Engineering Geology and Environment (IAEG).

List of Figures

List of Tables

Chapter 1
Introduction

1.1 Background

On January 13, 1863, the first underground railway system in the world—London Metropolitan Railway, was opened to traffic in London. The rapid development of the underground railway has brought urban transportation into the era of rail transit. At present, more than 300 subway lines have been opened in more than 100 cities around the world, with a total length of more than 7000 km, which makes the subway an important symbol of the economic development of a city. With the rapid development of economy and technology in China, the traffic pressure brought by the process of urbanization is very huge. Therefore, the construction of the underground railway is the most effective way to solve this problem. There are also many issues that need to be researched for geotechnical underground engineering.

Soft clay is widely distributed in China, especially in the southeast coastal areas. Since the end of the last century, with the rapid development of urban engineering construction in Shanghai, many large-scale transportation facilities and projects have been built on soft soil foundation. The bearing capacity of soft clay foundation is low, the strength increases slowly, and it has the characteristics of low permeability, high rheology, high water content and high compressibility, so its engineering properties are poor (Huang 1983). The problems such as urban ground subsidence caused by the mechanical properties of soft clay, excessive stratum deformation and instability of deep foundation pit excavation have become increasingly prominent.

After more than half a century of development, the subway construction methods in China have developed rapidly. At the beginning, the open cut method is mainly used, and now the surface excavation method, drilling and blasting method and shield method are adopted. No matter which construction method it is, the subway will inevitably produce vibration and settlement in the process of construction and operation, as well as the impact on the surrounding environment (Wang et al. 2003).

Z.-D. Cui et al., *Dynamics of Freezing-Thawing Soil around Subway Shield Tunnels*,
https://doi.org/10.1007/978-981-15-4342-5_1

Because of the advantages of small ground influence, fast progress and strong adaptability of geological conditions, shield method has been adopted in most subway construction projects in China, and has become the most advanced subway construction technology at present. In some complicated construction environments, especially in the southeast coastal areas, the conventional construction methods sometimes cannot maintain the stability of the soil around the tunnel, it is necessary to adopt some special construction methods, among which the freezing method is widely used because of its high strength, good controllability and strong applicability. In the process of freezing construction, the volume of pore water in the soil will change when it freezes, which increases the pore volume in the soil, leads to the expansion of the soil, the instability of the mechanical properties and the uneven surface deformation. And the soil is soft and easily deformed after thawing, producing greater uneven settlement. The effect of frost heaving and thaw collapsing is also the main problem of freezing method in the construction of subway tunnel (Yu et al. 2005).

During the operation of the subway, the vibration caused by the subway cyclic loading exists for a long time, which will reduce the strength of the soft soil foundation, generate different degrees of settlement and deformation, affect the deformation of the tunnel axis, and cause ground subsidence (Huang et al. 2006; Chen et al. 2005; Jiang 2002; Liu et al. 2004; Zhao 2006). The traffic facilities of many subway construction projects are usually built on soft soil foundation. In the initial stage of operation after the construction of freezing method, due to the action of cyclic loading, consolidation deformation settlement often occurs, which has great safety problems.

The settlement of a Japanese railway reached nearly 1 m in the five years after it was put into operation. There was almost no settlement on Shanghai Metro Line 1 before it was completed and opened to traffic, but its settlement reached 40–60 mm in eight months after operation. These examples indicate that soft clay will have large deformation and settlement under cyclic loading. If uneven settlement cannot be effectively controlled, it will cause tunnel leakage, segment fracture, building tilt or collapse, foundation instability and more serious land subsidence disasters.

Therefore, it is of great significance to study the dynamic characteristics of soil around subway shield tunnel by artificial freezing method, which can provide theoretical basis and guidance for the effective control of long-term subway settlement and maintaining the safety of subway operation.

1.2 Mechanical Properties of Soft Clay Under Cyclic Loading

Under cyclic loading, there are many factors affecting the dynamic characteristics of soft clay, including soil properties, loading frequency, cyclic stress ratio, vibration times, stress history, consolidation method, and drainage conditions. By using of indoor cyclic triaxial test, resonant column test, dynamic load test and computer numerical simulation, the dynamic characteristic parameters such as strength, pore water pressure, strain and elastic modulus of soft clay under cyclic load can be obtained.

Seed and Chan (1961) carried out the dynamic strength test of saturated soft clay and found that the soil has additional deformation after cyclic loading, which is mainly determined by consolidation pressure, dynamic stress amplitude and vibration times. Lo (1969) defined the concept of pore pressure ratio, and studied the relationship between the pore water pressure ratio and the soft clay strain, and considered that the relationship between them is not related to the cyclic loading time and consolidation time and consolidation pressure. Lee analyzed the residual strain of soft clay under cyclic loading, and fitted the empirical formula of residual strain after N times of cyclic loading, as shown in Eq. (1.1).

$$\varepsilon_p = 10\left\{\frac{\left[C_4' + S_4'(K_C - 1) + C_5'\sigma_3 + S_5'(K_C - 1)\sigma_3\right]^{-1}\sigma_d}{(0.1N)^{S_1}}\right\}^{C_6' + S_6'(K_C - 1)} \tag{1.1}$$

where K_c is consolidation ratio; σ_d is cyclic stress; C_4', C_5', C_6', S_4', S_5', S_6' and S_1 are the test parameters, changing with the soil properties.

Sangrey et al. (1969) performed a relatively low-frequency cyclic triaxial test to study the relationship between stress, strain, and pore pressure of soft clay under dynamic loading. Yasuhara et al. (1982) proposed that the normalized pore water pressure and shear strain have a unique hyperbolic relationship. Matsui et al. (1980) studied the dynamic characteristics of over-consolidated soft clay and normal consolidated remolded soft clay, and analyzed the variation of pore water pressure. Larew and Leonards (1962) first proposed the concept of critical cyclic stress ratio. When the cyclic stress is lower than the critical cyclic stress, the soft clay will not be destroyed under cyclic loading. When the cyclic stress is higher than the critical cyclic stress, the soft clay will fail. This concept has been confirmed by many researchers in subsequent studies. Yasuhara et al. (1982) and Yasuhara (1985) also pointed out that there is a hyperbolic function relation between pore pressure and shear strain, and proposed a hyperbolic function relationship of normalized pore pressure under cyclic loading, as shown in Eq. (1.2).

$$\Delta_u = \frac{\varepsilon}{\mathrm{a} + \mathrm{b}\varepsilon} \tag{1.2}$$

where Δ_u is cumulative residual pore pressure, ε is axial strain, a and b are test constants.

Hyodo et al. (1992) studied the dynamic characteristics of soft clay under cyclic loading by controlling the conditions of drainage and partial drainage. Mohanty et al. (2010) researched the strength and deformation characteristics of silty clay under cyclic loading through stress-controlled cyclic triaxial tests. Pan and Pande (1984) adopted a function similar to the form of excited load, which reflected the vibration characteristics of subway loads, as shown in Eq. (1.3).

$$F(t) = A_0 + A_1 \sin \omega_1 t + A_2 \sin \omega_2 t + A_3 \sin \omega_3 t \quad (1.3)$$

where A_0 is train weight; A_1, A_2, and A_3 are the dynamic stress amplitudes at the corresponding frequencies.

Zhou (1998) confirmed the concept of critical cyclic stress ratio proposed by Larew and pointed out that the critical cyclic stress ratio of Hangzhou soft clay is about 0.5. At the same time, the concept of threshold cyclic stress ratio was also proposed, which is the minimum cyclic stress ratio to make the soil produce excess pore water pressure and cumulative strain. Jiang and Chen (2001) found that the axial strain and body strain of soft clay have similar development trend under different waveforms, but the development rate is different. As the number of vibrations increases, the deformation rate generated by rectangular waves is the largest, and then triangular wave and tooth wave, while the deformation rate under the sine wave is the smallest. Zhang et al. (2010) presented the characteristics of the dynamic stress-strain curve of soft clay, and proposed a dynamic backbone curve model of soft clay. The influencing factors of the model include dynamic stress amplitude, vibration frequency, and consolidation pressure. Wang et al. (2008a, b) established a pore pressure-softening index model for over-consolidated soft clay, which reflects the variation of the soil softening index. The larger the cyclic stress ratio and the number of vibrations, the smaller the softening index. Jiang et al. (2009) studied the effect of cyclic stress ratio on the axial strain of soft clay by isostatic consolidation undrained static triaxial and dynamic triaxial tests. Tang et al. (2004) found that the dynamic strength of muddy silty clay decreased with the increase of cyclic load cycles under the vibration load of subway. Ding et al. (2015) studied the effects of consolidation degree and number of vibrations on the strain and pore pressure of saturated soft clay under drained and undrained conditions, and suggested that drainage should be considered in the settlement prediction of foundation soil after construction. Tang et al. (2008) pointed out that the effective principal stress attenuation of soft clay under subway vibration load has three stages. Yang and Liu (2016) believed that the magnitude and frequency of the cyclic load of the subway had a significant impact on the long-term settlement of the weak underlying layer of the tunnel. Ding et al. (2017) found that the fitting effect of bias sine wave on train cyclic load is better. Under the condition of partial drainage, the lower the degree of consolidation of soft clay is, the greater the peak value of pore water pressure is.

1.3 Physical and Mechanical Properties of Soft Clay Under Freeze-Thaw Cycle

In the process of subway construction, due to the complex construction environment in the soft soil area and the unstable soil around the tunnel, the artificial freezing method is often used to strengthen the soil. This method is not limited by the scope and depth of support, and is widely used in mine construction and subway connection channel construction. When the temperature is below 0 °C, the water in the soil pores will form ice and the volume will increase, which will lead to the increase of the pore volume and the occurrence of frost heave. During thawing, the ice melts into water, which leads to the decrease of soil pore volume and porosity, and the downward settlement of soil, causing the occurrence of thaw settlement. Frost heave and thaw settlement will destroy the internal structure and engineering properties of soil, which is a problem often encountered in the construction process of artificial freezing method.

After the soft clay is frozen, its physical and mechanical properties will be changed and its strength will be improved (Ling et al. 2013). Edwin and Anthony (1979) found that after freezing-thawing, the percentage of pores in the soft clay decreased and the permeability increased. Eigenbord (1996) also studied the permeability of soil under the action of freeze-thaw cycle, and pointed out that the void ratio of soil decreased, while the vertical permeability showed an increasing trend. Kim and Daniel (1992) found that the hydraulic conductivity of soil increased obviously after freezing-thawing, but different compaction methods had no obvious effect on the hydraulic conductivity, so it was considered that the porosity and structure of soil changed under the action of freeze-thaw cycle. Viklander (1998) found that the void ratio of high density soil will increase and the compactness will decrease after freeze-thaw cycle. Yang (2001) studied the difference of physical indexes between undisturbed soil and frozen-thawed soil. It was found that after freezing-thawing, the dry density and plasticity index of soil decreased slightly, while void ratio, liquid index and void ratio increased slightly. Wang et al. (2017) pointed out that with the increase of the number of freeze-thaw cycles, the frost heave rate of soft soil volume gradually increased, and the frost-heaving and thawing-settlement of the first freeze-thaw cycle was the most obvious. As shown in Fig. 1.1, the compression tests were conducted for the undisturbed soil and the freezing-thawing soil under different freezing temperatures and the lower the freezing temperature is, the greater the compressibility of soil is (Wang et al. 2009).

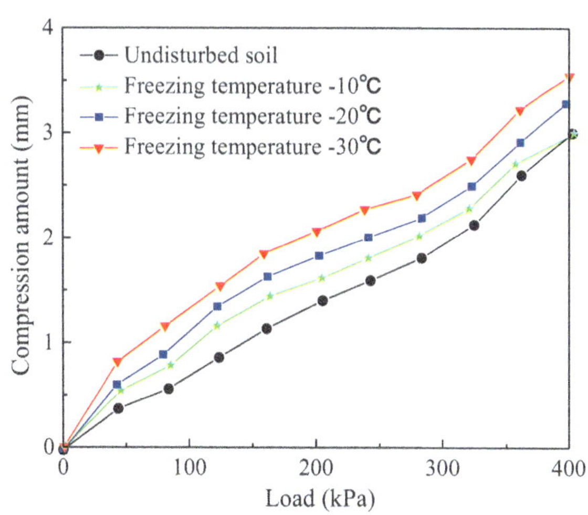

Fig. 1.1 Load-compression relation curve under different freezing temperatures

After freezing and thawing, the soft clay will produce micro-cracks and pores, the connection mode and structural shape of the soil particles will also be changed, and the cementation between the soil particles will be destroyed, which reduce the structural strength and change the mechanical properties of the soil (Qi et al. 2005). Figure 1.2 shows the development of axial strain of freeze-thaw soft clay under subway loading, and Fig. 1.3 indicates the variation of excess pore water pressure under freezing-thawing cycle at different frequencies.

Qi et al. (2006) found that whether it is loose soil or fine-grained soil, the soil strength will decrease after freeze-thaw cycles. Graham and Au (1985) carried out one-dimensional compression tests on the undisturbed clay after freeze-thaw cycles, and found that the soil structure was severely damaged after 5 freeze-thaw cycles. Leroueil et al. (1991) found that the peak shear stress of over-consolidated

Fig. 1.2 Development of axial strain under freezing-thawing cycle

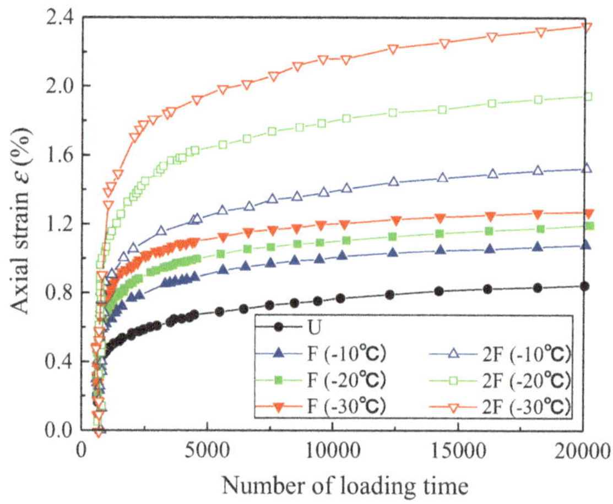

Fig. 1.3 Variation of excess pore water pressure under freezing-thawing cycle at different frequencies

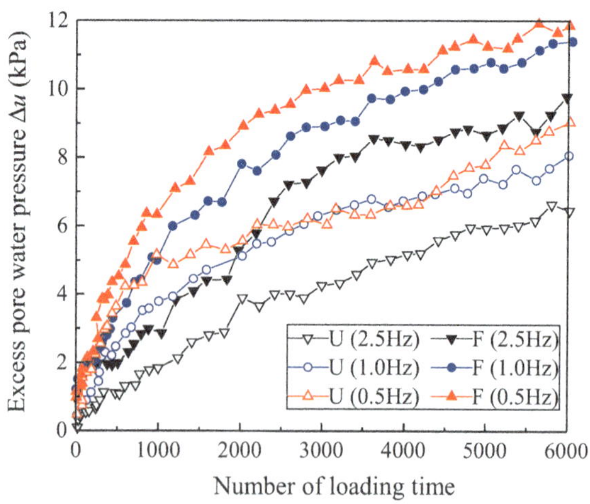

clay decreased obviously or even did not exist after freezing and thawing. The elastic modulus of frozen-thawed soil will also attenuate, which is related to the soil properties (Simonsen et al. 2002). Wang et al. (2010) studied the static and dynamic characteristics of cement soil under freeze-thaw cycle, and proved that the cyclic freeze-thaw action destroyed the internal structure of the soil, and the lower the freezing temperature was, the smaller the stiffness of the soil was. After freezing and thawing, the internal friction angle of soft clay increases and the cohesive force decreases, and the lower the freezing temperature, the greater the change in the internal friction angle (Wang et al. 2009). Xu et al. (2016) pointed out that the freeze-thaw cycle has a great influence on the dynamic properties of expansive soil, and attention should be paid to the damage effect of the first freeze-thaw cycle on expansive soil in practical engineering.

1.4 Macroscopic and Microstructural Characteristics of Soft Clay

As shown in Fig.1.4 for the macroscopic and microstructural characteristics of soil, we can not only study the internal structural morphology of soil, but also analyze the influence of internal microstructure on the physical and mechanical properties of soil, which plays an important role in establishing the relationship between microstructure and macro-mechanical properties of soft clay (Zhang 2007).

Nagaraj et al. (1998) performed compression tests on reinforced sensitive soft soil, established a corresponding microstructure model, and analyzed the development law of soil pore radius and permeability during compression. Shi (1996) summarized

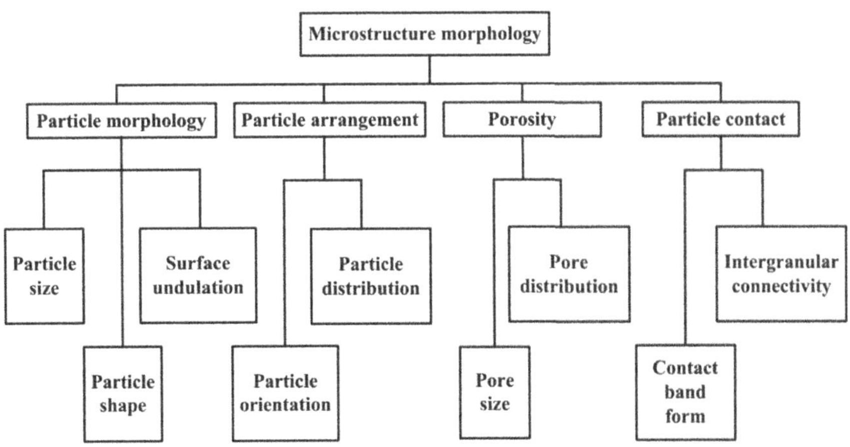

Fig. 1.4 Conceptual model of soil microstructure model system

the research results of soil microstructure and established a quantitative analysis method of soft clay pore structure parameters-SQM method. Tang et al. (2012) carried out scanning electron microscope tests on Shanghai muddy soft clay before and after freezing-thawing. It was found that the porosity and pore diameter of soft clay were greatly affected by freezing temperature. Wang et al. (2004, 2008a, b) studied the effects of threshold size and magnification on the quantitative analysis of soil microstructure, and found that the pore structure in the range of threshold [60, 100] and [150, 220] is more in line with the reality. Parameters such as pore ratio, pore direction, pore size and fractal dimension are all affected by cyclic stress ratio. The greater the cyclic stress ratio, the greater the degree of soil failure (Cao et al. 2014). Xu et al. (2015) established a three-dimensional pore calculation model of soft clay and quantitatively analyzed the microstructure by using the image processing technology of Image Pro Plus software.

Many researches also use mercury intrusion porosimetry to obtain soil pore morphology and distribution. Jiang et al. (2010) obtained a method for pore division of soil after dynamic load through MIP tests, and pointed out that the method of calculating the fractal dimension of pores under dynamic load by using Menger sponge model is feasible. Zhou et al. (2010) found that the pores of soft clay are mainly distributed in small pores, and the pore volume decreases with the increase of consolidation pressure. Regarding the study of soil microstructure under freeze-thaw cycle, Tang et al. (2005) analyzed the pore size, shape and contact state of Shanghai soft clay by SEM tests, and found that the soil pores became significantly larger after freeze-thaw cycle. Combining with the SEM tests and the MIP tests results, Zhang et al. (2015) pointed out that the freeze-thaw cycle changed the pore distribution of the soil. The pore size, number and diameter increased with the number of freeze-thaw cycles. Ding et al. (2016) found that honeycomb structure appeared in soft clay after freezing-thawing, the number of micro-pores decreased, and that of the macro-pores increased. The fractal dimension of pores increased, and the microstructure of soil changed more and more significantly with the decrease of freezing temperature and the increase of freezing and thawing cycles.

References

Cao, Y., Zhou, J., & Yan, J. J. (2014). Study of microstructures of soft clay under dynamic loading considering effect of cyclic stress ratio and frequency. *Rock and Soil Mechanics, 35*(3), 735–743.

Chen, Y. P., Huang, B., & Chen, Y. M. (2005). Deformation and strength of structural soft clay under cyclic loading. *Chinese Journal of Geotechnical Engineering, 27*(9), 1065–1071.

Ding, Z., Fan, J. Y., Zhang, M. Y., & Wei, X. J. (2017). Experimental study on pore pressure and strain model of undisturbed soil under subway dynamic loading. *Journal of the China Railway Society, 39*(3), 98–102.

Ding, Z., Zhang, M. Y., Wei, X. J., Hong, Q. H., & Zheng, Y. (2016). Experimental research on the microstructure of thawed soil after the subway construction of freezing method. *Journal of Railway Engineering Society, 33*(11), 106–112.

Ding, Z., Zhang, T., Wei, X. J., & Zhang, M. Y. (2015). Experimental study on effect of different drainage conditions on dynamic characteristics of soft clay under different degrees of consolidation. *Chinese Journal of Geotechnical Engineering, 37*(5), 893–899.

Edwin, J. C., & Anthony, J. G. (1979). Effect of freezing and thawing on the permeability and structure of soils. *Engineering Geology, 13,* 73–92.

Eigenbord, K. D. (1996). Effects of cyclic freezing and thawing on volume changes and permeabilities of soft fine-grained soils. *Canadian Geotechnical Journal, 33*(4), 529–537.

Graham, J., & Au, V. C. S. (1985). Effects of freeze-thaw and softening on a natural clay at low stresses. *Canadian Geotechnical Journal, 22*(1), 69–78.

Huang, W. X. (1983). *Engineering properties of soil.* Beijing: China Water & Power Press.

Huang, M. S., Li, J. J., & Li, X. Z. (2006). Cumulative deformation behaviour of soft clay in cyclic undrained tests. *Chinese Journal of Geotechnical Engineering, 28*(7), 891–895.

Hyodo, M., Yasuhara, K., & Hirao, K. (1992). Prediction of clay behavior in undrained and partially drained cyclic triaxial tests. *Soils and Foundation, 32*(4), 117–127.

Jiang, J. (2002). Study on the strain rate of clay under cyclic loading. *Chinese Journal of Geotechnical Engineering, 24*(4), 528–531.

Jiang, M. M., Cai, Z. Y., Cao, P., & Fang, W. (2009). Effects of cyclic loading on mechanical properties of silty clay. *Rock and Soil Mechanics, 30*(5), 205–207.

Jiang, J., & Chen, L. Z. (2001). One-dimensional settlement due to long-term cyclic loading. *Chinese Journal of Geotechnical Engineering, 2*(3), 366–369.

Jiang, Y., Lei, H. Y., Zheng, G., & Yang, X. J. (2010). Fractal study of microstructure variation of structured clays under dynamic loading. *Rock and Soil Mechanics, 31*(10), 3075–3080.

Kim, W., & Daniel, D. E. (1992). Effects of freezing on hydraulic conductivity of compacted clay. *Journal of Geotechnical engineering, 118*(7), 1083–1097.

Larew, H. G., & Leonards, G. A. (1962). A strength criterion for repeated loads. *Highway Research Board Proceedings, 41,* 529–556.

Leroueil, S., Tardif, J., Roy, M., La Rochelle, P., & Konrad, J. M. (1991). Effects of frost on the mechanical behaviour of Champlain Sea clays. *Canadian Geotechnical Journal, 28*(5), 690–697.

Ling, X. Z., Li, Q. L., Wang, L. N., Zhang, F., An, L. S., & Xu, P. J. (2013). Stiffness and damping radio evolution of frozen clays under long-term low level repeated cyclic loading: Experimental evidence and evolution model. *Cold Regions Science and Technology, 86,* 45–54.

Liu, Z. W., Qin, C. R., & Wang, J. H. (2004). Research on accumulative deformation behavior of soft clay under cyclic loads. *Water Resources and Hydropower Engineering, 35*(11), 14–17.

Lo KY (1969) The pore pressure-strain relationship of normally consolidated undisturbed clays: PART II. Experimental investigation and practical applications. Canadian Geotechnical Journal 6(4): 383–412.

Matsui, T., Ohara, H., & Ito, T. J. (1980). Cyclic stress-strain history and shear characteristics of clay. *Geotech Engng Div ASCE, 106*(10), 1101–1120.

Mohanty, B., Patra, N. R., & Chandra, S. (2010). Cyclic triaxial behavior of pond ash. In *Geo Florida 2010: Advances in Analysis, Modeling and Design, Florida* (vol. 20, pp. 833–841).

Nagaraj, T. S., Pandian, N. S., & Narashimha Raju, P. S. R. (1998). Compressibility behavior of soft cemented soils. *Geotechnique, 48*(2), 281–287.

Pan, C. S., & Pande, G. N. (1984). Preliminary deterministic finite element study on a tunnel driven in loess subjected to train loading. *China Civil Engineering Journal, 17*(4), 19–28.

Qi, J. L., Vermeer, P. A., & Cheng, G. D. (2005). State-of-the-art of influence of freeze-thaw on engineering properties of soils. *Advances in Earth Science, 20*(8), 887–893.

Qi, J. L., Vermeer, P. A., & Cheng, G. D. (2006). A review of the influence of freeze-thaw cycles on soil geotechnical properties. *Permafrost and Periglacial Processes, 7*(3), 245–252.

Sangrey, D. A., Henkel, D. J., & Esrig, M. I. (1969). The effective stress response of a saturated clay soil to repeated loading. *Canadian Geotechnical Journal, 6*(3), 241–252.

Seed, H. B., & Chan, C. K. (1961). Effect of duration of stress application on soil deformation under repeated loading. In *Proceedings of 5th International Congress on Soil Mechanics and Foundations, Pairs* (vol. 1, pp. 341–345).

Shi, B. (1996). Quantitative assessment of changes of microstructure for clayey soil the process of compaction. *Chinese Journal of Geotechnical Engineering, 18*(4), 57–62.

Simonsen, E., Janoo, V. C., & Isacsson, U. (2002). Resilient properties of unbound road materials during seasonal frost conditions. *Journal of Cold Regions Engineering, 16*(1), 28–50.

Tang, Y. Q., Shen, F., Hu, X. D., Zhou, N. Q., Zou, C. Z., & Zhu, J. H. (2005). Study on dynamic constitutive relation and microstructure of melted dark green silty soil in Shanghai. *Chinese Journal of Geotechnical Engineering, 27*(11), 1249–1252.

Tang, Y. Q., Wang, Y. L., Huang, Y., & Zhou, Z. Y. (2004). Dynamic strength and dynamic stress-strain relation of silt soil under traffic loading. *Journal of Tongji University (Natural Science), 32*(6), 701–704.

Tang, Y. Q., Zhao, S. K., Yang, P., Zhang, X., Wang, J. X., & Zhou, N. Q. (2008). Characteristic of deformation of saturated soft clay due to subway load. *Journal of Engineering Geology 16*(Sup): 54–60.

Tang, Y. Q., Zhou, J., Hong, J., Yang, P., & Wang, J. X. (2012). Quantitative analysis of the microstructure of Shanghai muddy clay before and after freezing. *Bulletin of Engineering Geology and the Environment, 71*(2), 309–316.

Viklander, P. (1998). Permeability and volume changes in till due to cyclic freeze-thaw. *Canadian Geotechnical Journal, 35*(3), 471–477.

Wang, J., Cai, Y. Q., & Li, X. B. (2008a). Cyclic softening-pore pressure generation model for over-consolidated clay under cyclic loading. *Rock and Soil Mechanics, 29*(12), 3217–3222.

Wang, P., Guo, W. J., Li, H. F., & Lu, H. J. (2017). Effect of cyclic freezing and thawing on physical and mechanical properties of soft soil. *Journal of Yangtze River Scientific Research Institute, 34*, 1–5.

Wang, C. J., Ji, M. X., & Chen, Y. M. (2003). Additional settlement of saturated soft clay foundation under train loading. In *The 9th Chinese Academy of Civil Engineering Geotechnical Engineering Conference*. Tsinghua University Press.

Wang, T. L., Liu, J. K., Peng, L. Y., & Tian, Y. H. (2010). Research on the mechanical properties of the cement-modified soil under the action of freezing and thawing cycles. *China Railway Science, 31*(6), 7–13.

Wang, B. J., Shi, B., Cai, Y., & Tang, C. S. (2008b). 3D visualization and porosity computation of clay soil SEM image by GIS. *Rock and Soil Mechanics, 29*(1), 251–255.

Wang, B. J., Shi, B., Liu, Z. B., & Cai, Y. (2004). Fractal study on microstructure of clayey soil by GIS. *Chinese Journal of Geotechnical Engineering, 26*(2), 244–247.

Wang, X. B., Yang, P., Wang, H. B., & Dai, H. M. (2009). Experimental study on effects of freezing and thawing on mechanical properties of clay. *Chinese Journal of Geotechnical Engineering, 31*(11), 1768–1772.

Xu, R. Q., Deng, Y. W., Zhan, X. G., Xu, L. Y., & Lu, J. Y. (2015). Soft soil three-dimensional porosity calculated based on SEM image and its influence factors analysis. *Chinese Journal of Rock Mechanics and Engineering, 34*(7), 1497–1502.

Xu, L., Liu, S. H., Lu, Y., Song, Y. J., & Yang, Q. (2016). Physico-mechanical properties of expansive soil under freeze-thaw cycles. *Rock and Soil Mechanics, 37*(2), 168–173.

Yang, P. (2001). Study on the difference of physical and mechanical properties between original and thawing soil. *Journal of Nanjing Forestry University (Natural Sciences Edition), 25*(2), 68–70.

Yang, B. M., & Liu, B. G. (2016). Analysis of long-term settlement of shield tunnel in soft soil area under cyclic loading of subway train. *China Railway Science, 37*(3), 62–66.

Yasuhara, K. (1985). Undrained and drained cyclic triaxial tests on a marine clay. In *Proceedings of the 11th ICSMFE* (vol. 2, pp. 1095–1098).

Yasuhara, K., Yamanouchi, T., & Hirao, K. (1982). Cyclic strength and deformation of normally consolidated clay. *Soils and Foundations, 22*(3), 77–91.

Yu, Z. K., Huang, H. W., & Wang, R. L. (2005). Application of the artificially ground freezing method to shanghai metro engineering. *Journal of Glaciology and Geocryology, 27*(2), 550–552.

Zhang, X. (2007). *Study on the micro-structure and dynamic characteristic of soft clay around tunnel under the subway-included loading*. Ph.D. thesis, Tongji University.

Zhang, Y., Bing, H., & Yang, C. S. (2015). Influences of freeze-thaw cycles on mechanical properties of silty clay based on SEM and MIP test. *Chinese Journal of Rock Mechanics and Engineering, 34*(1), 3598–3602.

Zhang, Y., Kong, L. W., & Li, X. W. (2010). Dynamic backbone curve model of saturated soft clay under cyclic loading. *Rock and Soil Mechanics, 31*(6), 1699–1708.

Zhao, S. K. (2006). *The Study on the Micro-structure Distortion Mechanics of Soft Clay Under the Subway-Included Loading*. Master degree thesis, Tongji University.

Zhou, J. (1998). *Study on the characteristics of saturated soft clay under cyclic loading*. Ph.D. thesis, Zhejiang University.

Zhou, H., Fang, Y. G., & Zeng, C. (2010). Application of mercury intrusion porosimetry in the study of pore distribution of saturated soft soil. *Yellow River, 32*(5), 101–102.

Chapter 2
Artificial Freezing for by-Pass of Subway Tunnel

2.1 Artificial Freezing Method

Artificial freezing method is a special construction method for geotechnical engineering, which freezes the water in the soil around the underground excavation space into ice with artificial refrigeration technology. The soil and ice will be cemented together to form a freezing wall or a closed underground space, which can isolate the groundwater and bear the earth pressure. So the underground construction can be carried out under the protection of the freezing wall. In practical, the ammonia-brine refrigeration technology is used in most of the projects. The ammonia-brine system consists of brine circulation system, Freon circulation system and cooling water circulation system.

The construction of the artificial freezing method is as follows. Firstly, drill freezing holes and lay freezing pipes in the stratum. Secondly, install the refrigeration system and let the low temperature brine circulate in the freezing pipes. Thirdly, the terrestrial heat is taken away and the water in the soil is frozen into ice. Finally, a closed underground space will be formed. The physical and mechanical properties of the frozen soil are much better than those of soft soil. So the freezing wall can form a closed underground space with certain strength to bear the earth pressure and water pressure.

When the thickness and average temperature of the freezing wall reach the design requirements, the excavation of the underground space can be carried out under the protection of the freezing wall. During the excavation process, the inside of the freezing wall is exposed directly to the air, which results in the loss of cooling capacity. Therefore, in order to ensure the safety of the excavation, the refrigeration system shouldn't stop until the secondary lining is completed. The development of the frozen soil is shown in Fig. 2.1 (Jin et al. 2008).

There are two artificial freezing methods for reinforcing the soil, low temperature brine method and liquid nitrogen method. Compared with the low temperature brine method, the liquid nitrogen method characterized by low temperature, high speed

Z.-D. Cui et al., *Dynamics of Freezing-Thawing Soil around Subway Shield Tunnels*,
https://doi.org/10.1007/978-981-15-4342-5_2

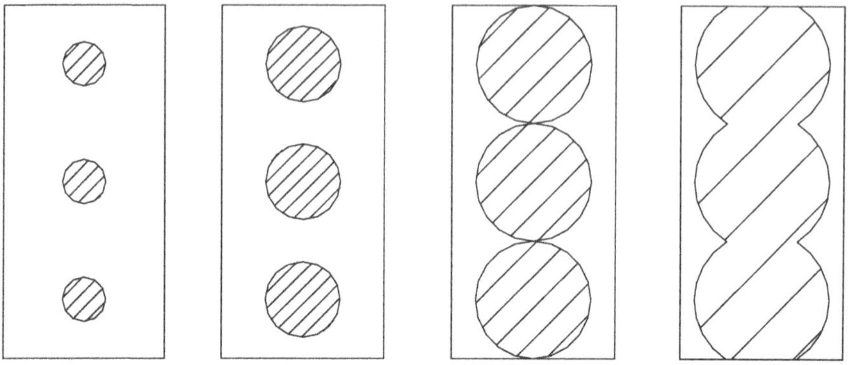

(a) Freezing pipes (b) Development of the frozen cylinders (c) Closure of the freezing wall (d) Formation of the freezing wall

Fig. 2.1 Schematic diagram of the development of the frozen soil

and high strength is suitable for rescue engineering. However, the low temperature brine method characterized by relatively low cost is suitable for projects with huge works and long construction period, which is widely used in the subway by-pass and shield tunnel portal soil reinforcement engineering.

2.2 Design of Freezing Parameters of by-Pass

2.2.1 Project Overview

There is usually a by-pass with a pump station in the subway tunnel. The horizontal centerline elevation of the upline tunnel is −16.217 m and that of the downline tunnel is −6.213 m at the location of the by-pass. The top of the by-pass is about 16.4 m away from the ground surface. The by-pass consists of a bell-mouthed opening, a corridor, and a pump station, as is shown in Fig. 2.2 (Zhan et al. 2020).

The cross section shape of the by-pass is straight-wall-top-arch whose primary lining is wood lagging, steel support and shotcrete of C25. The thickness of the primary lining is 300 mm. Besides, the secondary lining is the waterproof reinforced concrete of C35P10, whose thickness is 450 mm. The height of the excavation outline of the by-pass and the pump station is 7.833 m and its width is 3.320 m. The height of the excavation outline of the bell-mouthed opening is 4.283 m and its width is 3.920 m.

The soil information of each layer from top to bottom at the location of the by-pass is summarized in Table 2.1. Figure 2.3 shows that the soil around the by-pass mainly

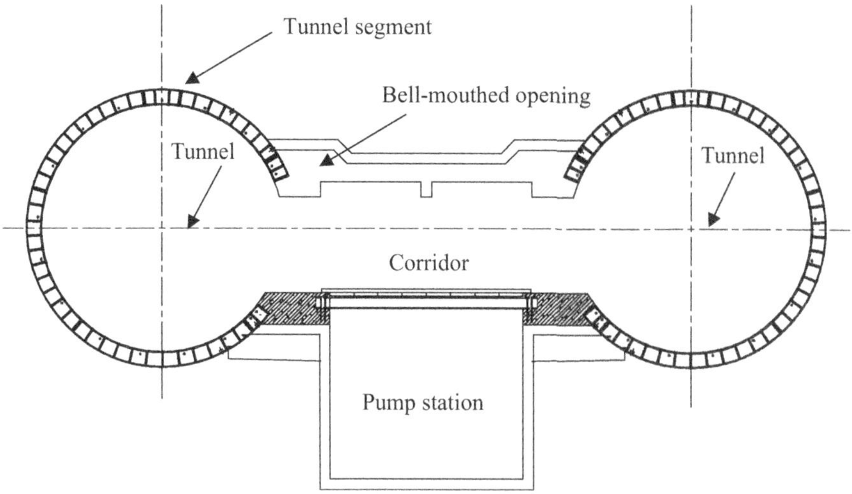

Fig. 2.2 Schematic diagram of a by-pass

Table 2.1 Physical and mechanical indexes of different soil layers

Soil layer	Unit weight (kN/m^3)	Average thickness (m)	Coefficient of static earth pressure K_0
Filling soil ①	20	1.184	/
Clay ③$_2$	19.5	4.840	0.45
Silty clay ③$_3$	18.9	0.480	0.48
Silty sand mixed with silt ⑤$_1$	19.2	6.264	0.35
Silty sand ⑤$_2$	19.2	4.412	0.32
Silty clay ⑥$_3$	19.5	9.528	0.45

includes silty sand and silty clay, which are characterized by high water content, high compressibility, low strength and low permeability.

The weighting unit weight of the soil above the by-pass is

$$\begin{aligned}\gamma_0 &= \sum_{i=1}^{5}(\gamma_i h_i)/\sum_{i=1}^{5} h_i\\ &=(20\times 1.184+19.5\times 4.840+18.9\times 0.48+19.2\times 6.264+19.2\times 3.632)/16.4\\ &=19.3\,\text{kN/m}^3\end{aligned}$$

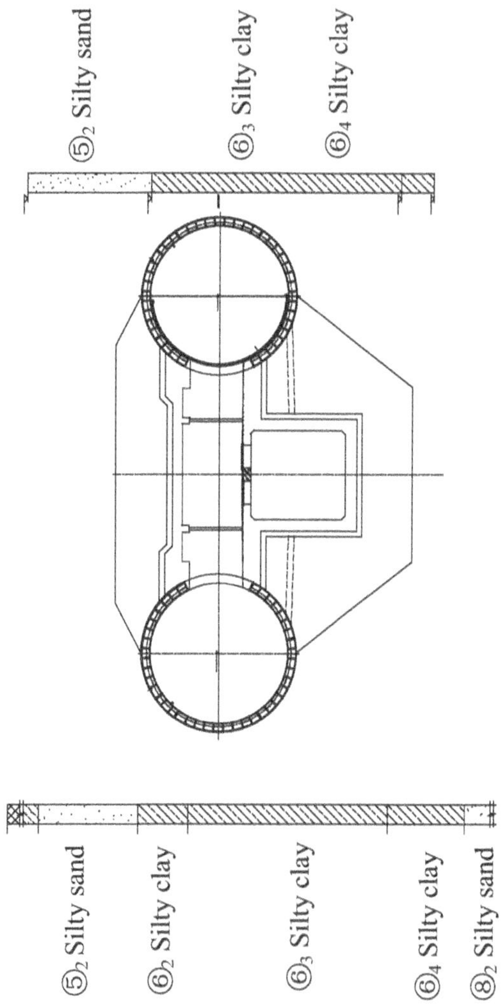

Fig. 2.3 Stratigraphic profile of the by-pass

2.2.2 Design of Thickness of the Freezing Wall

2.2.2.1 Numerical Method

An example of a by-pass is as follows. The force model of the freezing wall is shown in Fig. 2.4. The thickness of the freezing wall is 1.8 m. Figure 2.4a shows the force model of the freezing wall outside the corridor. The width of the outer contour is 6.92 m and its height is 7.562 m. The width of the inner contour is 3.32 m and its height is 3.962 m. The inner diameter of the circular arch is 2.41 m and its outer diameter is 4.21 m. Figure 2.4b shows the force model of the freezing wall outside the pump station. The width of the outer contour is 6.92 m and its height is 5.65 m. The width of the inner contour is 3.32 m and its height is 3.85 m. The frozen soil is ⑥$_3$ silty clay and its burial depth is 16.4 m. The lateral pressure coefficient is 0.7. The value of the temporary load is 20 kN/m^2, including the dead weight of buildings and the ground overload within the construction area. When the average temperature of the frozen soil is −10 °C, its strength indices and main parameters are summarized in Table 2.2.

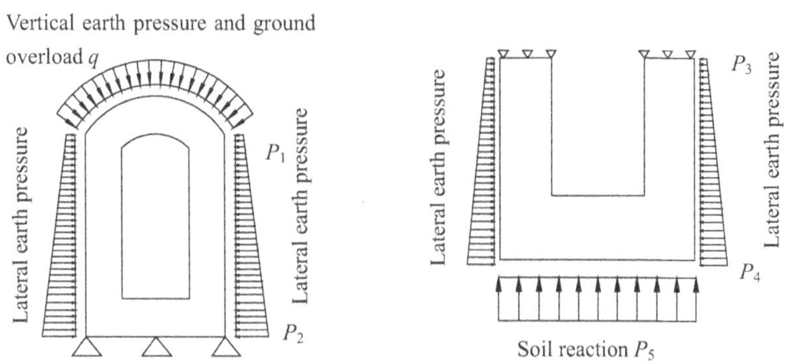

(a) The force model of the freezing wall outside the corridor

(b) The force model of the freezing wall outside the pump station

Fig. 2.4 The force model of the freezing wall

Table 2.2 Strength indices and main parameters of the frozen soil for −10 °C

Compressive strength $[f_c]$ (MPa)	Shear strength $[\tau]$ (MPa)	Flexural strength $[f_t]$ (MPa)	Modulus of elasticity of the frozen soil (MPa)	Poisson's ratio of the frozen soil μ	Cohesive strength (MPa)	Friction angle (°)
3.50	1.50	2.00	177.33	0.264	1.50	10

1) Loads acting on the freezing wall outside the corridor

(1) Vertical earth pressure and ground overload q

$$q = 19.3\,\mathrm{kN/m^3} \times 16.400\,\mathrm{m} + 20\,\mathrm{kN/m^2} = 0.337\,\mathrm{MPa}$$

(2) Lateral earth pressure

$$P_1 = 19.3\,\mathrm{kN/m^3} \times 16.400\,\mathrm{m} \times 0.7 = 0.222\,\mathrm{MPa}$$
$$P_2 = 19.3\,\mathrm{kN/m^3} \times 22.152\,\mathrm{m} \times 0.7 = 0.299\,\mathrm{MPa}$$

2) Loads acting on the freezing wall outside the pump station

(1) Soil reaction

$$P_5 = 19.3\,\mathrm{kN/m^3} \times 26.013\,\mathrm{m} + 20\,\mathrm{kN/m^2} = 0.522\,\mathrm{MPa}$$

(2) Lateral earth pressure

$$P_3 = 19.3\,\mathrm{kN/m^3} \times 20.383\,\mathrm{m} \times 0.7 = 0.275\,\mathrm{MPa}$$
$$P_4 = 19.3\,\mathrm{kN/m^3} \times 26.013\,\mathrm{m} \times 0.7 = 0.351\,\mathrm{MPa}$$

The analysis of the internal force and deformation of the freezing wall is carried out by numerical simulation. The computational model of the freezing wall outside the corridor is shown in Fig. 2.5. And the internal force and deformation are shown in Fig. 2.6 and Table 2.3.

Under the action of loads mentioned above, the cross-sectional stress of the freezing wall should meet the following requirements.

$$\sigma \leq \frac{[\sigma]}{K} \tag{2.1}$$

where σ represents the cross-sectional stress (MPa); $[\sigma]$ represents the compressive strength, flexural strength or shear strength of the frozen soil (MPa) and K is the safety coefficient.

From Tables 2.3 and 2.4, we can obtain that the compressive safety coefficient of the freezing wall at the inner corner is 1.78, which is smaller than 2. It means that stress concentration happens at the inner corner of the freezing wall. However, the calculation value is still smaller than the compressive strength of the frozen soil and circular arcs are often used instead of right angles in practical engineering. Therefore, it is safe. In addition, the internal force of the freezing wall meets the requirements listed in Table 2.4.

The computational model of the freezing wall outside the pump station is shown in Fig. 2.7. And the internal force and deformation are shown in Fig. 2.8.

Fig. 2.5 The finite element model of the freezing wall outside the corridor

From Fig. 2.8, we can obtain that the maximum compressive stress of the freezing wall outside the pump station happens at the inner corner along the long side, which is 1.44 MPa. The maximum tensile stress happens on the inside of the floor, which is 0.13 MPa. The maximum shear stress happens at the inner corner along the long side, which is 0.53 MPa. The maximum displacement happens at the floor, which is 27.6 mm. Therefore, the bearing capacity and stability of the freezing wall outside the pump station meets the requirements listed in Tables 2.2 and 2.4.

2.2.2.2 Force-Method

The computational model of the corridor is shown in Fig. 2.9. f = 1.222 m, L = 2.560 m, h = 4.526 m, $\alpha = 51^\circ = 0.89$ rad, R = 3.294 m, e = P_1 = 0.222 MPa, $\Delta e = P_2 - P_1 = 0.077$ MPa.

According to the principle of force-method in structural mechanics (Zheng 2011), the calculation formula is

$$\left.\begin{aligned}\delta_{11}x_1 + \delta_{12}x_2 + \Delta_{1p} = 0\\ \delta_{21}x_1 + \delta_{22}x_2 + \Delta_{2p} = 0\end{aligned}\right\} \tag{2.2}$$

where x_1 and x_2 are calculated with diagrammatic multiplication method. Figures 2.10 and 2.11 illustrate the bending moment diagram of the basic structure under the action of unit force and outer force, respectively.

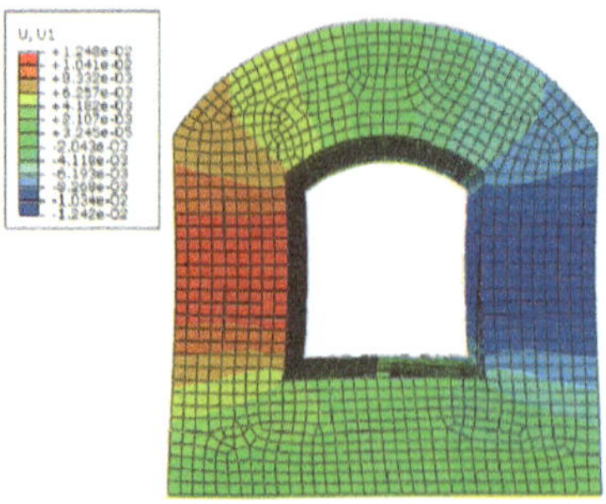

(b) Distribution of U_x

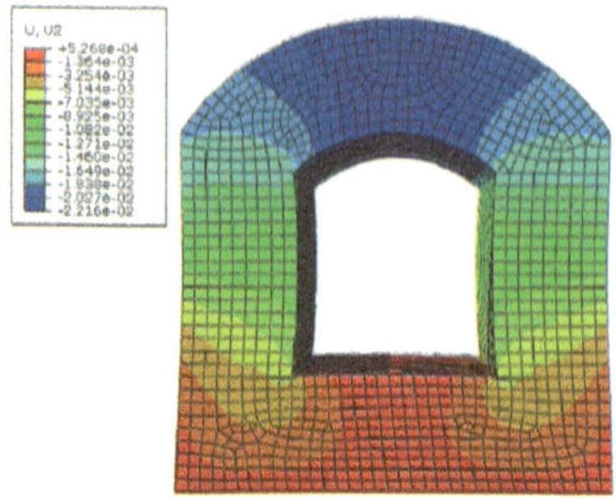

(c) Distribution of U_y

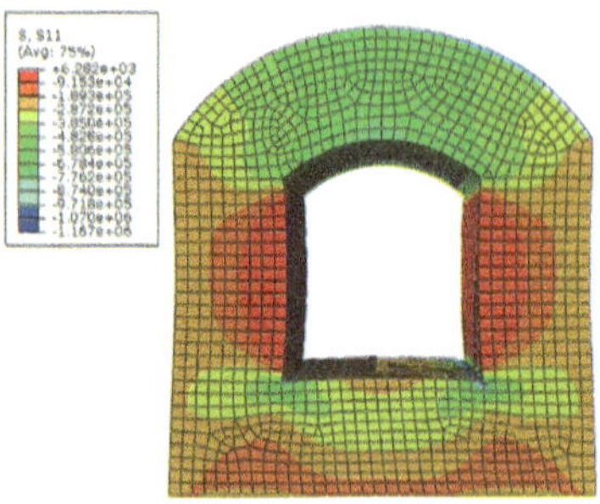

(d) Distribution of σ_x

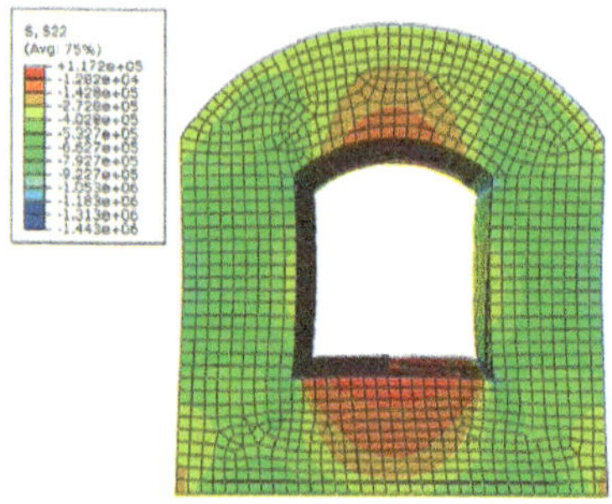

(e) Distribution of σ_y

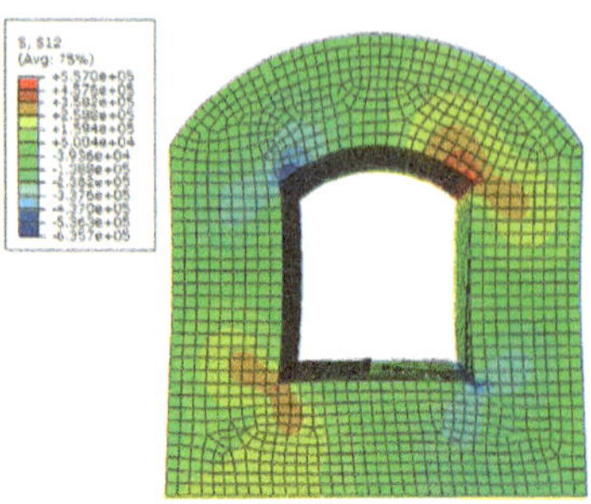

(f) Distribution of τ_{xy}

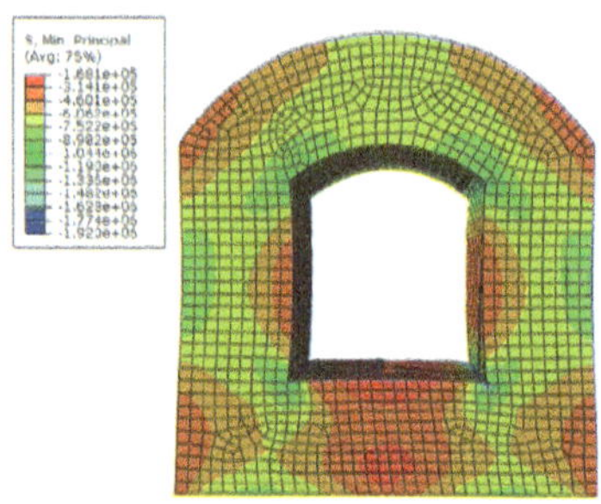

(g) Distribution of minimum principal stress

Fig. 2.6 The finite element analysis model and analysis result of the freezing wall outside the corridor

$$\begin{aligned}\delta_{11} &= \frac{1}{EI}\int_0^{\alpha} R^2(1-\cos\theta)^2 d\theta + \frac{h}{6EI}\left[2f^2 + 2(f+h)^2 + 2f(f+h)\right] \\ &= \frac{R^2}{EI}\left(\frac{3\alpha}{2} + \frac{\sin 2\alpha}{4} - 2\sin\alpha\right) + \frac{h}{3EI}\left[3f^2 + h^2 + 3fh\right] \\ &= \frac{3.294^2}{EI}\left(\frac{3}{2}\times 0.89 + \frac{\sin(2\times 0.89)}{4} - 2\sin 0.89\right)\end{aligned}$$

Table 2.3 The internal force and deformation of the freezing wall outside the corridor

Indices	Compressive strength (MPa)					Shear strength (MPa)	Displacement (mm)	
Position	Inside of the side wall	Outside of the side wall	Inside of the arch	Outside of the arch	Conner	Conner	Vault	Middle of the side wall
Calculation value	0.32	0.96	0.67	0.64	1.92	0.64	22.2	12.4
Strength of the frozen soil	3.50	3.50	3.50	3.50	3.50	1.50		
Safety coefficient	10.94	3.65	5.22	5.47	1.82	2.34		
Notes					Stress concentration			

Table 2.4 The safety coefficient of the freezing wall of class III

Indices	Compressive safety coefficient K_1	Flexural safety coefficient K_2	Shear safety coefficient K_3
Safety coefficient	2.0	3.0	2.0

Fig. 2.7 The finite element model of the freezing wall outside the pump station

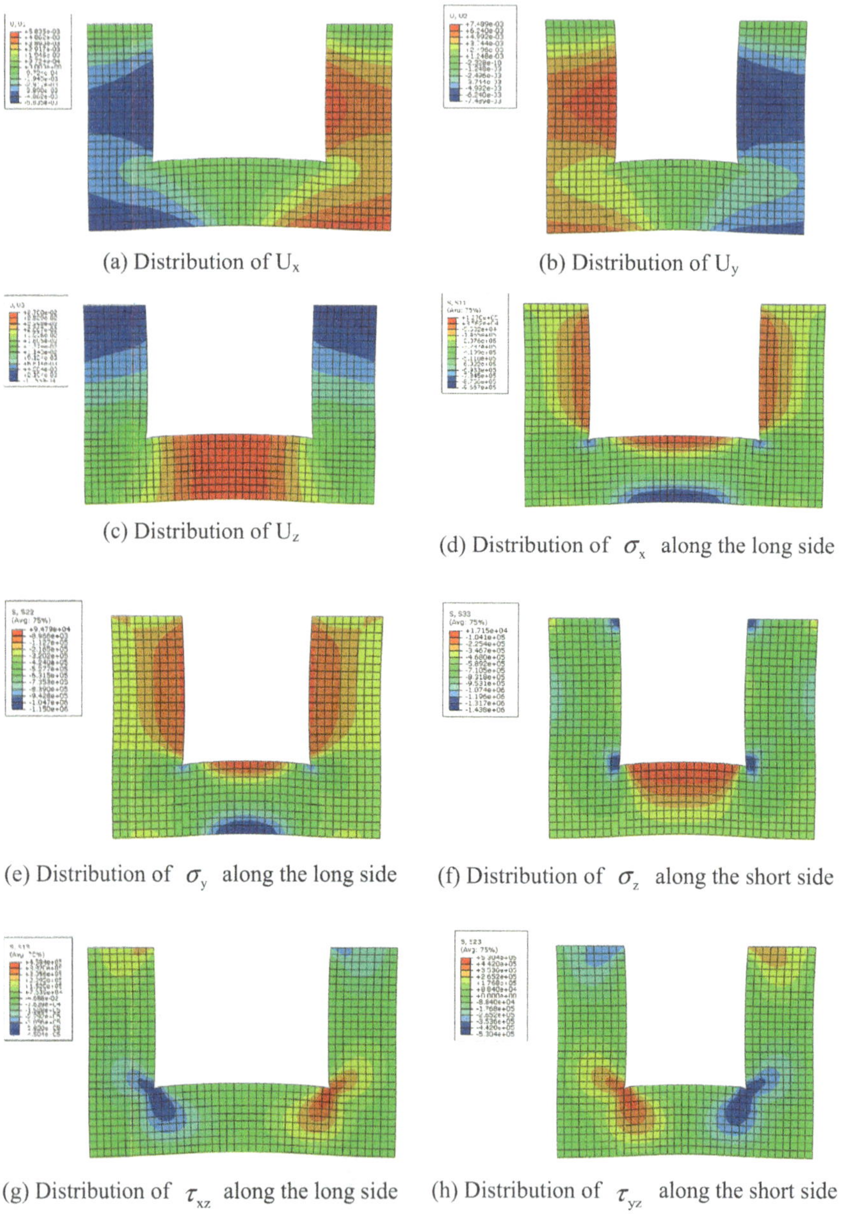

(a) Distribution of U_x

(b) Distribution of U_y

(c) Distribution of U_z

(d) Distribution of σ_x along the long side

(e) Distribution of σ_y along the long side

(f) Distribution of σ_z along the short side

(g) Distribution of τ_{xz} along the long side

(h) Distribution of τ_{yz} along the short side

Fig. 2.8 The finite element analysis model and analysis result of the freezing wall outside the pump station

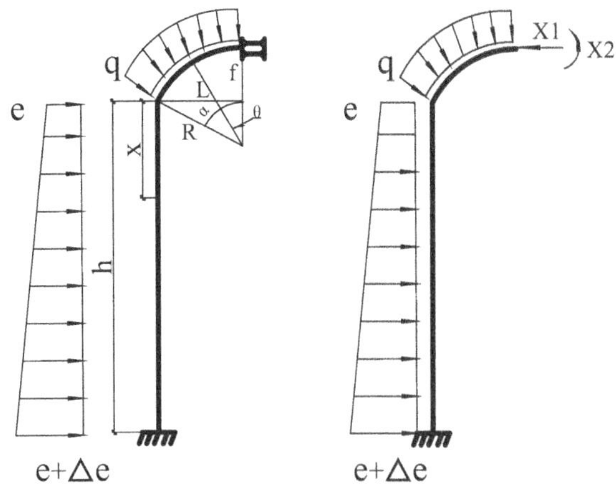

Fig. 2.9 The computational model of the freezing wall

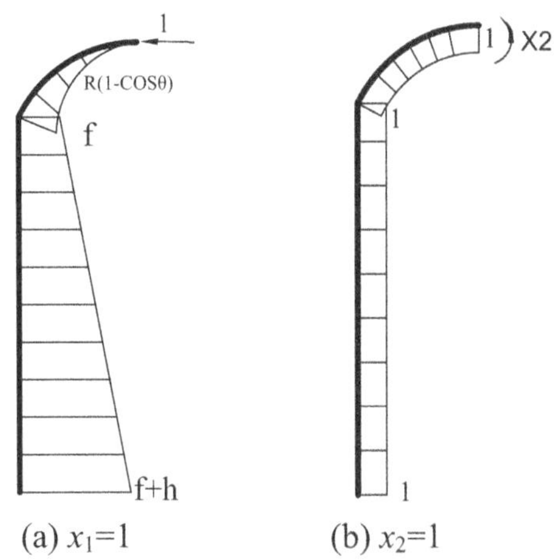

Fig. 2.10 The bending moment diagram of the basic structure under the action of unit force

$$
\begin{aligned}
&+\frac{4.526}{3EI}\left[3\times 1.222^2+4.526^2+3\times 1.222\times 4.526\right]\\
&=\frac{62.97}{EI}
\end{aligned}
$$

$$
\delta_{22}=\frac{\alpha R+h}{EI}=\frac{0.89\times 3.294+4.526}{EI}=\frac{7.46}{EI}
$$

$$
\begin{aligned}
\delta_{12}=\delta_{21}&=\frac{1}{EI}\int_0^{\alpha}R^2(1-\cos\theta)d\theta+\frac{h}{EI}\left(\frac{2f+h}{2}\right)\\
&=\frac{1}{EI}\left(\alpha R^2-R^2\sin\alpha+fh+\frac{h^2}{2}\right)
\end{aligned}
$$

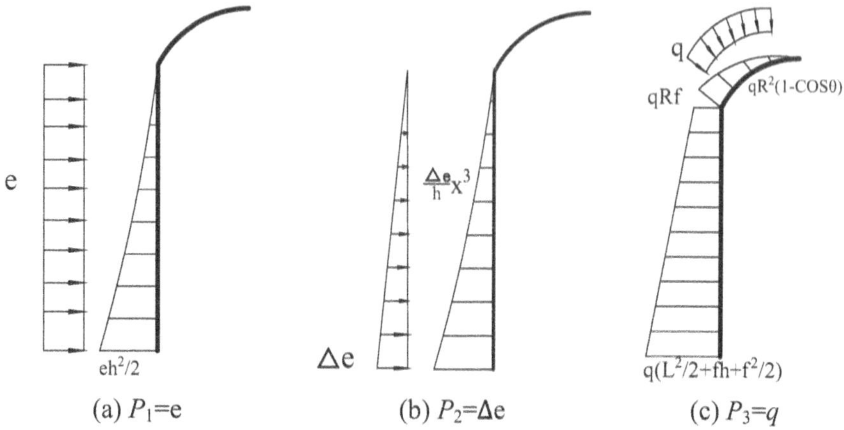

Fig. 2.11 The bending moment diagram of the basic structure under the outer force

$$
\begin{aligned}
&= \frac{1}{EI}\left(0.89 \times 3.294^2 - 3.294^2 \times \sin 0.89 + 1.222 \times 4.526 + \frac{4.526^2}{2}\right) \\
&= \frac{17.00}{EI}
\end{aligned}
$$

$$
\begin{aligned}
\Delta_{2p_1} &= -\frac{1}{EI}\left(\frac{1}{3} \times \frac{eh^2}{2} \times h\right) = -\frac{eh^3}{6EI} \\
&= -\frac{222000 \times 4.526^3}{6EI} = -\frac{3430404.812}{EI}
\end{aligned}
$$

$$
\begin{aligned}
\Delta_{2p_2} &= -\frac{1}{EI}\left(\frac{1}{4} \times h \times \frac{\Delta e}{h} \times h^3\right) = -\frac{\Delta e h^3}{4EI} \\
&= -\frac{77000 \times 4.526^3}{4EI} = -\frac{1784737.639}{EI}
\end{aligned}
$$

$$
\begin{aligned}
\Delta_{1p_3} &= -\frac{1}{EI}\left\{\begin{aligned} &\int_0^{\alpha} R(1-\cos\theta)qR^2(1-\cos\theta)Rd\theta + \frac{1}{2}qRfh\left[\frac{2}{3}f + \frac{1}{3}(f+h)\right] \\ &+\frac{1}{2}qh\left(\frac{L^2}{2} + fh + \frac{f^2}{2}\right)\left[\frac{1}{3}f + \frac{2}{3}(f+h)\right] \end{aligned}\right\} \\
&= -\frac{1}{EI}\left[\begin{aligned} &qR^4\left(\frac{3\alpha}{2} + \frac{1}{4}\sin 2\alpha - 2\sin\alpha\right) + \frac{1}{2}qRf^2h + \frac{1}{6}qRh^2f \\ &+\frac{1}{4}q\left(\frac{2L^2h^2}{3} + fhL^2 + hf^3 + \frac{4fh^3}{3} + \frac{8h^2f^2}{3}\right) \end{aligned}\right]
\end{aligned}
$$

$$= -\frac{1}{EI}\left[\begin{array}{r} 337000 \times 3.294^4 \times \left(\frac{3 \times 0.89}{2} + \frac{1}{4}\sin(2 \times 0.89) - 2\sin 0.89\right) \\ +\frac{1}{2} \times 337000 \times 3.294 \times 1.222^2 \times 4.526 + \frac{1}{6} \times 337000 \times 3.294 \times 4.526^2 \times 1.222 \\ +\frac{1}{4} \times 337000 \times \left(\begin{array}{r} \frac{2 \times 2.56^2 \times 4.526^2}{3} + 1.222 \times 4.526 \times 2.56^2 + 4.526 \times 1.222^3 \\ +\frac{4 \times 1.222 \times 4.526^3}{3} + \frac{8 \times 4.526^2 \times 1.222^2}{3} \end{array}\right) \end{array}\right]$$

$$= -\frac{36899570.19}{EI}$$

$$\Delta_{2\mathrm{p}_1} = -\frac{1}{EI}\left(\frac{1}{3} \times \frac{eh^2}{2} \times h\right) = -\frac{eh^3}{6EI}$$

$$= -\frac{222000 \times 4.526^3}{6EI} = -\frac{3430404.812}{EI}$$

$$\Delta_{2\mathrm{p}_2} = -\frac{1}{EI}\left(\frac{1}{4} \times h \times \frac{\Delta e}{h} \times h^3\right) = -\frac{\Delta eh^3}{4EI}$$

$$= -\frac{77000 \times 4.526^3}{4EI} = -\frac{1784737.639}{EI}$$

$$\Delta_{2p_3} = -\frac{1}{EI}\left\{\int_0^\alpha qR^2(1 - \cos\theta)d\theta + \frac{1}{2}h\left[qRf + q\left(\frac{L^2}{2} + fh + \frac{f^2}{2}\right)\right]\right\}$$

$$= -\frac{1}{EI}\left[qR^2(\alpha - \sin\alpha) + \frac{1}{4}(qhL^2 + 2qfh^2 + qhf^2 + 2qRfh)\right]$$

$$= -\frac{1}{EI}\left[\begin{array}{r} 337000 \times 3.294^2 \times (0.89 - \sin 0.89) \\ +\frac{1}{4}\left(\begin{array}{r} 337000 \times 4.526 \times 2.56^2 + 2 \times 337000 \times 1.222 \times 4.526^2 \\ +337000 \times 4.526 \times 1.222^2 + 2 \times 337000 \times 3.294 \times 1.222 \times 4.526 \end{array}\right) \end{array}\right]$$

$$= -\frac{10769226.85}{EI}$$

$$\Delta_{1\mathrm{p}} = \Delta_{1\mathrm{p}_1} + \Delta_{1\mathrm{p}_2} + \Delta_{1\mathrm{p}_3} = -\frac{15836463.82}{EI} - \frac{8643127.437}{EI} - \frac{36899570.19}{EI}$$

$$= -\frac{61379161.45}{EI}$$

$$\Delta_{2\mathrm{p}} = \Delta_{2\mathrm{p}_1} + \Delta_{2\mathrm{p}_2} + \Delta_{2\mathrm{p}_3} = -\frac{3430404.812}{EI} - \frac{1784737.639}{EI} - \frac{10769226.85}{EI}$$

$$= -\frac{15984369.3}{EI}$$

Substituting $\delta_{11}, \delta_{22}, \delta_{12}, \delta_{21}, \Delta_{1p}, \Delta_{2p}$ into Eq. (2.2), we can obtain

$$x_1 = 1.033\,\mathrm{MPa},\; x_2 = -0.211\,\mathrm{MPa}$$

Therefore, the axial force F_{N}, bending moment M and shear force Q of the side wall are

$$\left.\begin{aligned} F_N &= qR\sin\alpha = 862692.6 \\ M &= (f+x)x_1 + x_2 - qR^2(1-\cos\alpha) - \frac{1}{2}ex^2 - \frac{\Delta e}{6h}x^3 \\ &= 1033000x - 304000 - 111000x^2 - 2835.47x^3,\ x \in (0,h) \\ Q &= x_1 + qR(1-\cos\alpha) - ex - \frac{\Delta ex^2}{2h} = 1444483.3 - 222000x - 8506.4x^2 \end{aligned}\right\} \tag{2.3}$$

$$|M|_{\max} = 1870.6\ \text{kN}\cdot\text{m},\ |Q|_{\max} = 1444.5\ \text{kN},\ |F_N|_{\max} = 862.7\ \text{kN}$$

The axial force F'_N, bending moment M' and shearing force Q' of the arch are

$$\left.\begin{aligned} F'_N &= x_1\cos\theta - qR(1-\cos\theta) = 77000\cos\theta - 1110000 \\ M' &= R(1-\cos\theta)x_1 + x_2 - qR^2(1-\cos\theta) = -254000(1-\cos\theta) - 211000,\ \theta \in (0,\alpha) \\ Q' &= x_1\sin\theta - qR(1-\cos\theta) = 1033000\sin\theta - 1110000(1-\cos\theta) \end{aligned}\right\} \tag{2.4}$$

$$|M'|_{max} = 305.2\ \text{kN}\cdot m,\ |Q'|_{max} = 406.3\text{kN},\ |F'_N|_{max} = 1061.5\ \text{kN}$$

The axial stress of the freezing wall is calculated by the following formula.

$$\sigma = \frac{|M|}{W} \pm \frac{|F_N|}{A} = \frac{|M|}{\frac{1\times b^2}{6}} \pm \frac{|F_N|}{1\times b} \tag{2.5}$$

where M represents the bending moment of the freezing wall (N·m); F_N represents the axial force of the freezing wall (N); A represents the area of the freezing wall per unit length (m^2) and W represents the section modulus of the freezing wall per unit length (m^3).

The shear stress of the freezing wall is calculated by the following formula.

$$\tau = 1.5\frac{Q}{A} = 1.5\frac{Q}{1\times b} \tag{2.6}$$

where Q represents the shearing force of the freezing wall (N).

From Eqs. (2.1), (2.5) and (2.6), the thickness of the side wall should meet the following requirements.

$$\left.\begin{aligned} b_1 &\gg \max\left[\frac{|F_N|K_1 + \sqrt{K_1^2F_N^2 + 24K_1[f_c]M}}{2[f_c]}\right] = \max\left[\frac{862.7\times 2 + \sqrt{2^2\times 862.7^2 + 24\times 2\times 3500\times 1870.6}}{2\times 3500}\right] = 2.8\,\text{m} \\ b_2 &\gg \max\left[\frac{-|F_N|K_2 + \sqrt{K_2^2F_N^2 + 24K_2[f_t]M}}{2[f_t]}\right] = \max\left[\frac{-862.7\times 3 + \sqrt{3^2\times 862.7^2 + 24\times 3\times 2000\times 1870.6}}{2\times 2000}\right] = 3.5\,\text{m} \\ b_3 &\gg \max\left[\frac{1.5K_3Q}{[\tau]}\right] = \max\left[\frac{1.5\times 2\times 1444.5}{1500}\right] = 2.9\,\text{m} \end{aligned}\right\} \tag{2.7}$$

where b_1, b_2 and b_3 are the minimum thickness of the side wall that need to meet the requirements of compressive strength, flexural strength and shear strength respectively; $[f_c]$, $[f_t]$ and $[\tau]$ are the compressive strength, flexural strength and shear strength of the frozen soil respectively and K_1, K_2 and K_3 are the safety coefficient of the compressive strength, flexural strength and shear strength respectively.

From Eqs. (2.1), (2.5) and (2.6), the thickness of the arch should meet the following requirements.

$$\left.\begin{aligned} b_4 &\gg \max\left[\frac{|F'_N|K_1+\sqrt{K_1^2F_N'^2+24K_1[f_c]|M'|}}{2[f_c]}\right]=\max\left[\frac{1061.5\times 2+\sqrt{2^2\times 1061.5^2+24\times 2\times 3500\times 305.2}}{2\times 3500}\right]=1.4\ \mathrm{m}\\ b_5 &\gg \max\left[\frac{-|F'_N|K_2+\sqrt{K_2^2F_N'^2+24K_2[f_t]|M'|}}{2[f_t]}\right]=\max\left[\frac{-1061.5\times 3+\sqrt{3^2\times 1061.5^2+24\times 3\times 2000\times 305.2}}{2\times 2000}\right]=1.1\ \mathrm{m}\\ b_6 &\gg \max\left[\frac{1.5K_3Q'}{[\tau]}\right]=\max\left[\frac{1.5\times 2\times 406.3}{1500}\right]=0.8\ \mathrm{m} \end{aligned}\right\} \tag{2.8}$$

where b_4, b_5 and b_6 are the minimum thickness of the arch that need to meet the requirements of compressive strength, flexural strength and shear strength respectively.

Substituting $M_{\max}$, $F_{N\max}$, $Q_{\max}$, $M'_{\max}$, $F'_{N\max}$, $Q'_{\max}$ into Eqs. (2.7) and (2.8), the design thickness B of the freezing wall is

$$B=\max(b_l,b_2,b_3,b_4,b_5,b_6)=\max(2.8,3.5,2.9,1.4,1.1,0.8)=3.5\,\mathrm{m}$$

Compared with the numerical result, the force-method is probably on the conservative side.

2.2.3 Design of Freezing Holes

The thickness and average temperature of the freezing wall, the brine temperature and the freezing time should be considered when determining the spacing of the freezing holes. According to Technical Code for By-pass Freezing Method in Shanghai (2006), the spacing of the single-row freezing holes can be determined by Table 2.5. Deviation precision of the freezing holes can be determined by Table 2.6.

Table 2.5 The reference spacing of the single-row freezing holes

Type of freezing holes	Horizontal or tilted freezing holes			Vertical freezing holes	
Length of freezing holes (m)	≤10	10–30	30–60	≤40	40–100
Spacing of freezing holes (mm)	1100–1300	1300–1600	1600–2000	1200–1400	1400–1800

Table 2.6 Deviation precision requirements of the freezing holes

Type of freezing holes	Horizontal or tilted freezing holes			Vertical freezing holes	
Length of freezing holes (m)	≤10	10–30	30–60	≤40	40–100
The maximum deviation of the freezing holes (mm)	150	150–350	350–600	150–250	250–400

The length of the freezing hole is

$$L_{ks} = L_{sj} + L_0 + L_1 \tag{2.9}$$

where L_{ks} represents the length of the freezing hole (mm); L_{sj} represents the distance between the orifice of the freezing hole and the design boundary of the freezing wall (mm); L_0 represents the end of the freezing hole that can't circulate brine, and L_0 shouldn't be more than 150 mm and L_1 represents the poor freezing length (mm).

The freezing holes are mainly arranged in one tunnel and distributed around the by-pass. Five rows of freezing pipes are attached along the inner side of the segments to weaken the influence of heat exchange caused by tunnel ventilation. And the spacing of two adjacent freezing pipes is 400 mm. The range of the five-row freezing pipes shouldn't be less than the thickness of the freezing wall. The range of the insulation layer on the tunnel surface should be 2 m larger than the design boundary of the freezing wall. The thickness of the insulation layer should be 30–50 mm. The freezing holes for reinforcing the pump station are densely arranged and tilted downward. The total number of the freezing holes is 74 and their lengths are all smaller than 10 m.

According to Tables 2.5 and 2.6, the distance of the freezing holes should be smaller than 1.3 m. The maximum allowable deviation is 150 mm. The arrangement of the freezing pipes is shown in Fig. 2.12. The low carbon seamless steel tubes whose inner diameter is 89 mm and thickness is 8 mm are used as freezing pipes. The compressive strength of the low carbon seamless steel pipes is not less than 0.8 MPa and not less than 1.5 times the brine pressure.

2.2.4 Design of Average Temperature of Freezing Wall

The average temperature of the freezing wall is determined by the burial depth of the by-pass, the influence of the frost heave and thawing settlement and the rationality of the technological process. The larger the bearing capacity of the freezing wall is, the lower the average temperature is. Generally, the average temperature is determined according to Table 2.7. The average temperature of the freezing wall is −10 °C because the burial depth of the by-pass is 16.4–26.0 m.

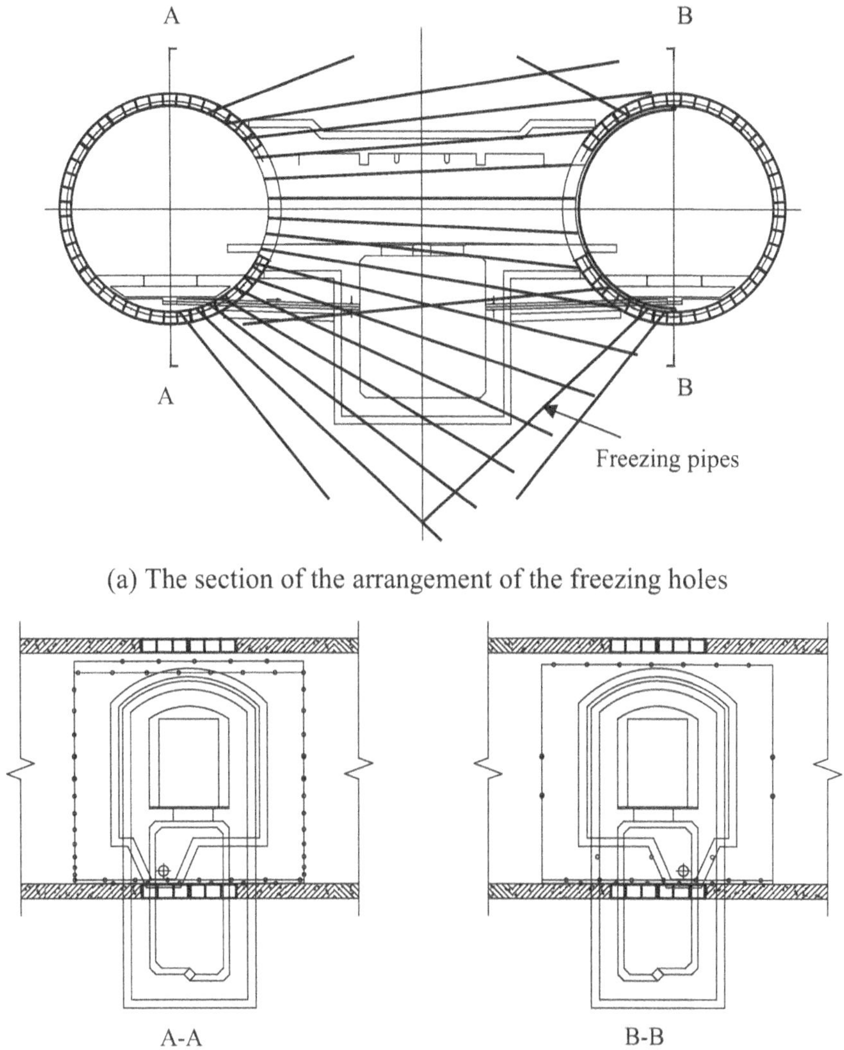

(a) The section of the arrangement of the freezing holes

(b) The elevation of the arrangement of the freezing holes

Fig. 2.12 Layout plan of the freezing holes

Table 2.7 The reference average temperature of the freezing wall

Burial depth H_j (m)	<12	12–30	>30
Average temperature T_p (°C)	−6 to −8	−8 to −10	≤−10

2.2.5 Design of Brine Temperature and Flow Rate

Figure 2.13 shows the predicted cooling curve of the brine temperature. The cooling speed should not be too fast, otherwise the freezing pipes will be cracked. The brine temperature should decrease to −18 °C for 7 days and −24 °C for 15 days. According to Tables 2.8 and 2.9, the lowest brine temperature shouldn't be higher than −28 °C for this by-pass. The brine temperature should be lower than −28 °C before excavation. The brine temperature shouldn't be higher than −28 °C until the structural construction is completed. In addition, the cyclic temperature difference of the brine temperature should not be higher than 2 °C and the brine flow of the single pipe shouldn't be smaller than 5 m^3/h.

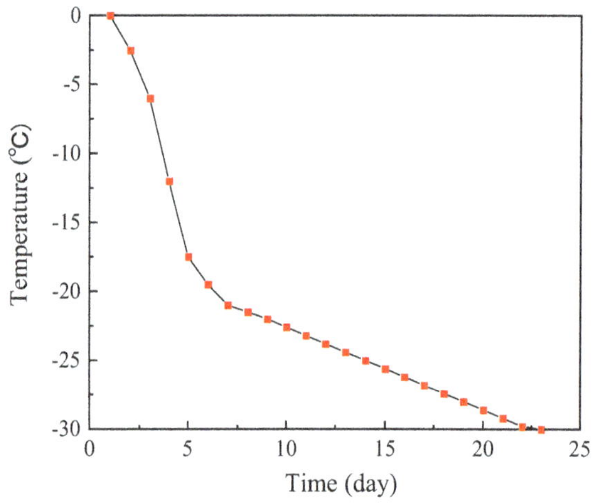

Fig. 2.13 Predicted cooling curve of the brine temperature

Table 2.8 Reference value of the minimum brine temperature

Average temperature T_p (°C)	−6 to −8	−8 to −10	≤−10
The minimum brine temperature T_p (°C)	−26 to −28	−28 to −30	−30 to −32

Table 2.9 Reference value of the brine flow of the single pipe

Total length of the freezing holes L_k (m)	≤40	40–80	>80
The minimum brine flow Q_{yk} (m^3/h)	3.0 to −5.0	5.0–8.0	≥8.0

2.3 Construction Technology of the Artificial Freezing Method

The construction method of by-pass includes open cut method, undercutting method and the combination of open cut and undercutting method. There are many undercutting methods for reinforcing soil such as deep-mixing pile or grouting method, pipe jacking method and freezing method. The open cut method is suitable for all kinds of soil characterized by simple construction, large construction area and great environmental influence. Deep-mixing pile or grouting method are suitable for cohesive soil characterized by barring traffic, mud or noise pollution and ground settlement. The pipe jacking method is not suitable for the soil with feeble confined water, and the displacement of the main tunnel will be caused by the poor control of the jacking force. Freezing method is suitable for any water bearing stratum characterized by good waterproof capability and bearing capacity, pollution-free and smooth traffic.

The freezing method combined with undercutting method is used, because the by-pass is located in the urban area and the stratum is a water bearing soft soil layer.

2.3.1 Procedures of the Freezing Construction

The construction procedures of the by-pass and the pump station are shown in Fig. 2.14. The key procedures are the construction of the freezing holes, the active freezing and the excavation of the by-pass.

Before the construction of the freezing holes, the backfill grouting shall be carried out for the ten-ring segments on both sides of the by-pass to improve the bottom soil and increase the structural stability of the surrounding segments, so as to reduce the risk of convergence and deformation during the construction of the by-pass.

During the active freezing period, monitoring of the brine circulation system and the soil temperature should be carried out to ensure that the temperature falls down normally. The thickness and average temperature of the freezing wall must meet the design requirements before excavation.

The full section excavation with the step distance between 0.5 and 0.8 m is adopted. The corridor is excavated firstly, then the bell-mouthed opening at both ends, and finally the pump station.

2.3.2 Technical Parameters of the Freezing Construction

2.3.2.1 Technical Parameters of the Freezing Hole Construction

Figure 2.15 shows the stand pipe that is used in the construction of freezing holes to avoid the water inflow and soil leakage. The hole position error of the freezing hole

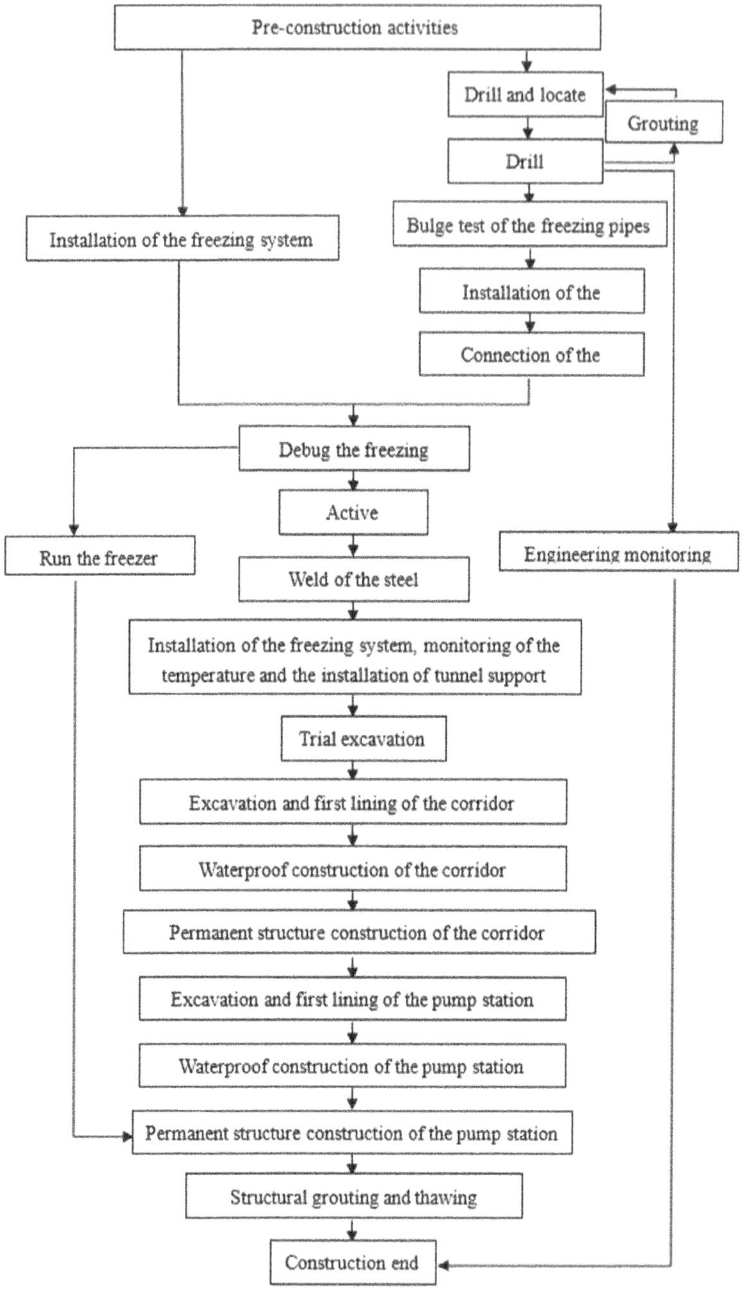

Fig. 2.14 The construction procedures of the by-pass and the pump station

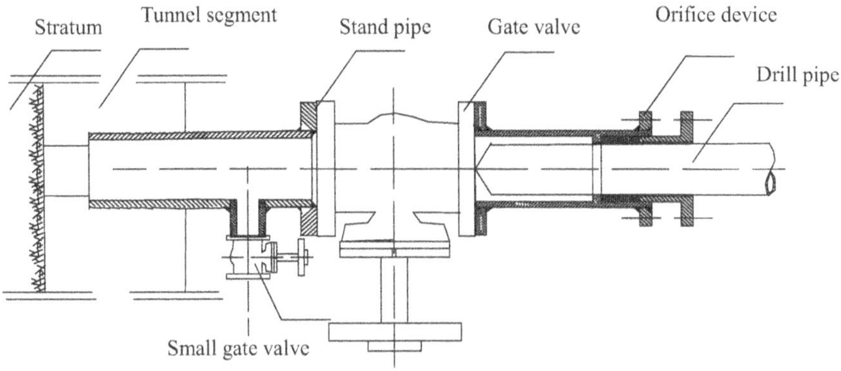

Fig. 2.15 Diagram of the hole sealing device

should be controlled within 100 mm. And the hole position should avoid the joints, bolts and main steel bars. For the first time, the hole depth is 280 mm. Get out the drilled core and install the stand pipe and the DN125 gate valve. Open the gate valve and carry out the secondary drill until the concrete segments are penetrated. The diameter of the hole is 108 mm. The DN125 gate valve must be closed immediately if the water inflow and soil leakage happens.

2.3.2.2 Technical Parameters of the Excavation

Before the excavation, the thickness of the freezing wall and the average temperature should be calculated according to the measured temperature data so as to ensure they reach the design requirements. In addition, the thickness and average temperature of the freezing wall should be verified to determine the freezing effect by drilling prospect holes in the poor freezing zone. The position of the prospect holes should near the design boundary of the freezing wall. In addition, it should be determined by the hole deviation diagram and measured temperature data. In principle, the position of the prospect holes should be located on the places characterized by relatively large end hole spacing, poor brine circulation and high measured temperature.

The brine temperature shouldn't be higher than −28 °C until the structural construction is completed. In addition, the cyclic temperature difference shouldn't be higher than 2 °C.

2.3.3 Design and Construction of the Emergency Door

2.3.3.1 Design of the Emergency Door

The function of the emergency door is to ensure the safety of the tunnel when the water inflow and soil leakage happen during the excavation of the by-pass.

The elevation of the tunnel centerline at the by-pass is about −14.353 m. The elevation of the ground surface is about +4.850 m. The top elevation of the emergency door is −13.252 m and the bottom elevation is −15.454 m. The lateral pressure coefficient k_0 is 0.7. The installation position of the emergency door is shown in Fig. 2.16 and its force model is shown in Fig. 2.17.

1) Loads acting on the emergency door

   At the top of the emergency door, the load is

$$Q_s = k_0 \gamma_0 H = 0.7 \times 19.3\,\text{kN/m}^3 \times (4.850\ \text{m} + 13.252\ \text{m}) = 0.244\ \text{MPa}$$

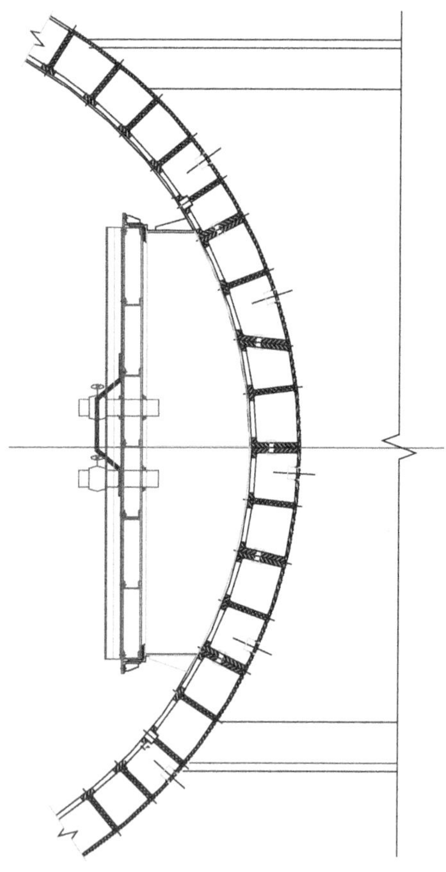

Fig. 2.16 The installation position of the emergency door

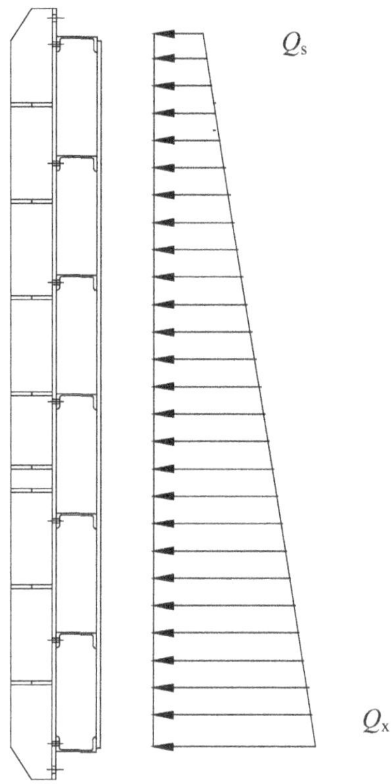

Fig. 2.17 The force model of the emergency door

At the bottom of the emergency door, the load is

$$Q_x = k_0 \gamma_0 H = 0.7 \times 19.3\ \text{kN/m}^3 \times (4.850\ \text{m} + 15.454\ \text{m}) = 0.274\ \text{MPa}$$

Therefore, the design load is

$$Q_t = (Q_s + Q_x)/2 = (0.244\ \text{MPa} + 0.274\ \text{MPa})/2 = 0.259\ \text{MPa}$$
$$Q = kQ_t = 1.3 \times 0.259\ \text{MPa} = 0.337\ \text{MPa}$$

where Q represents the calculation water and earth pressure (MPa); Q_t represents the design water and earth pressure (MPa); H represents the calculation depth (m); k_0 represents the lateral pressure coefficient and k represents the impact dynamic coefficient, which is 1.3.

2) Calculation of the emergency door and connectors

   (1) Strength check of the emergency door
   The structure diagram of the emergency door is shown in Fig. 2.18. It is a composite structure composed of 12 mm thick steel plates, 10 channel steels and 12 mm thick stiffened plates. Q235B steel is chosen and its

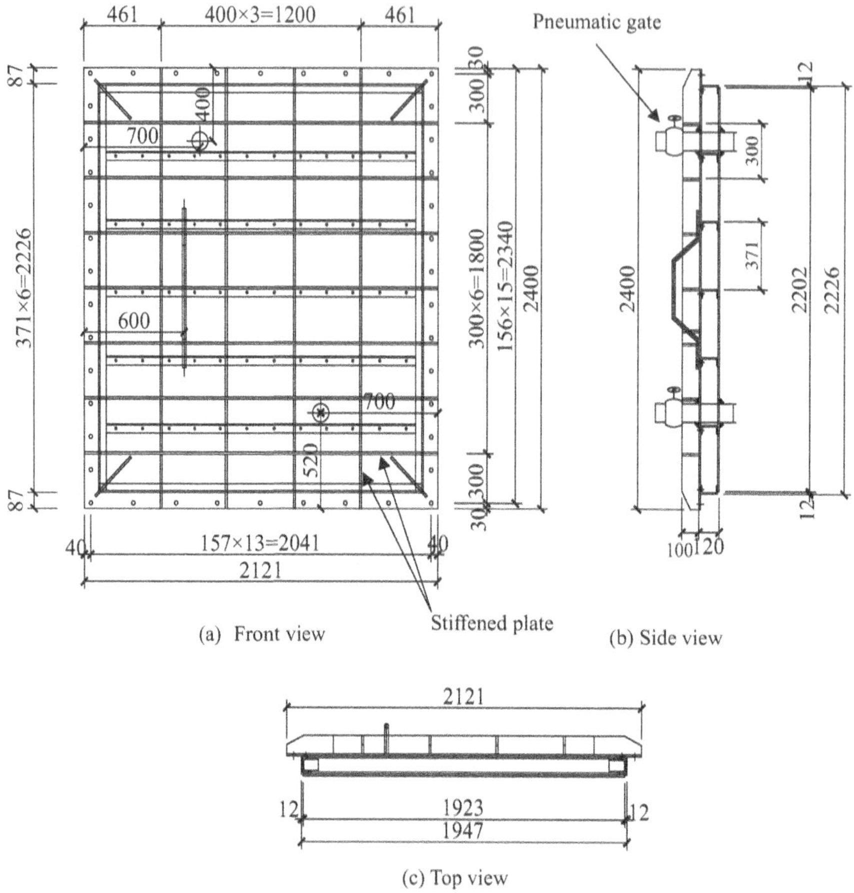

Fig. 2.18 The structure diagram of the emergency door

elastic modulus is 2.06×10^5 N/mm^2. Poisson's ratio is 0.25. Yield stress is 235 MPa. $L_x = 1.923$ m, $L_y = 2.202$ m. The analysis of the internal force and deformation of the emergency door is carried out by numerical simulation. The computational model is shown in Fig. 2.19. And the internal force and deformation are shown in Fig. 2.20.

From Fig. 2.20d, we can see that stress concentration happens at the end of the stiffened plates. The maximum stress is 296.6 MPa but the area of stress concentration is very small. From the side view and top view in Fig. 2.18, we can see that the ends of the stiffened plates are passivated to avoid the stress concentration in practical engineering. The stress of the other parts is not more than 155.9 MPa and the maximum shear stress is not more than 122.1 MPa, which are smaller than the yield strength of Q235B. So they meet the strength requirements.

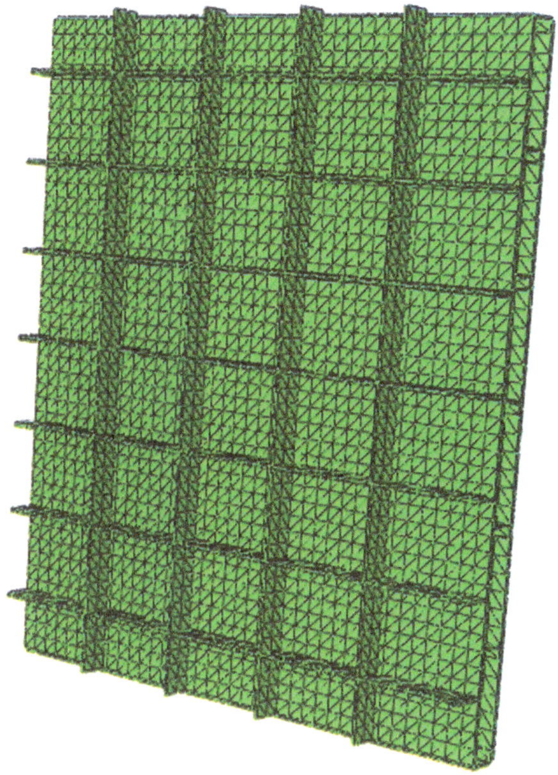

Fig. 2.19 The finite element model of the emergency door

From Fig. 2.20a, we can see that the maximum deformation appears at the center of the inside of the emergency door, which is 1.2 mm. According to Code for Design of Steel Structure (2014), the maximum deformation shouldn't be more than $L_y/150$, that is 14.7 mm. So it meets the requirement of deformation.

(2) Strength check of the bolts

The total load acting on the emergency door is

$$Q' = Q \times L_x \times L_y = 0.337\ \text{MPa} \times 1.923\ \text{m} \times 2.202\ \text{m} = 1427.0\ \text{kN}$$

According to Hexagon head bolts-Product grade C (GB/T5780-2000), M22 bolt is chosen, whose tensile strength is 170 N/mm^2 and effective sectional area is 3.03 cm^2. Therefore, the number of the bolts is

$$Q'' = 170\ \text{N/mm}^2 \times 3.03\ \text{cm}^2 = 51.51\ \text{kN}$$

$$n = \frac{Q''}{Q''} = \frac{1427.0\ \text{kN}}{51.51\ \text{kN}} = 28$$

So the number of the bolts is determined to be 42. There are 12 bolts on the long side and 9 on the short side.

(3) Strength calculation of the butt weld
Q235B steel is chosen for the door. Hand welding and E43 electrode are chosen. The inspection standard of the weld is grade three.

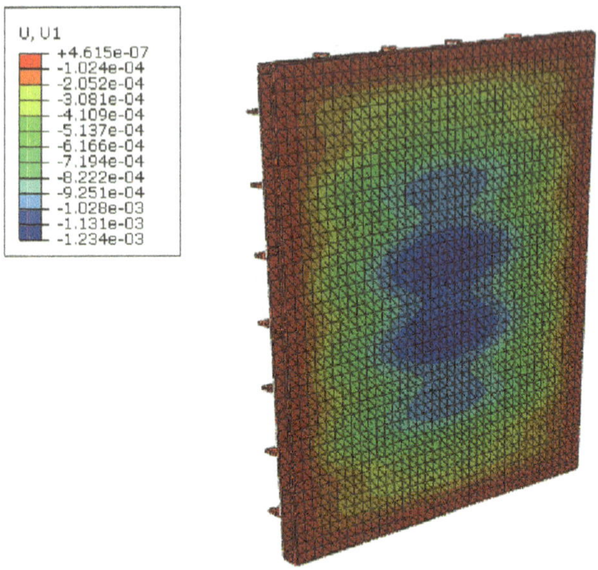

(a) Distribution of U_x

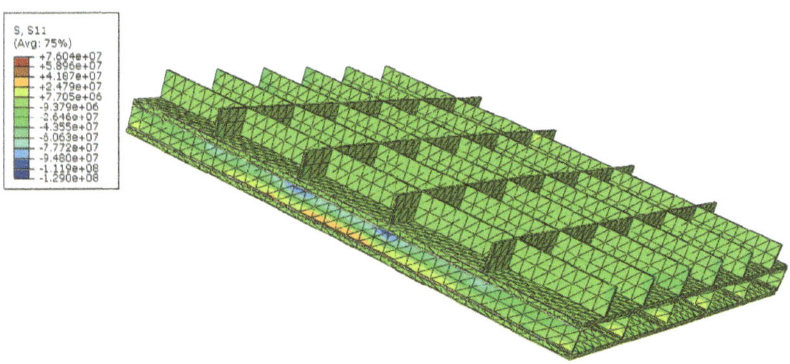

(b) Distribution of σ_x

Fig. 2.20 The finite element analysis model and analysis result of the emergency door

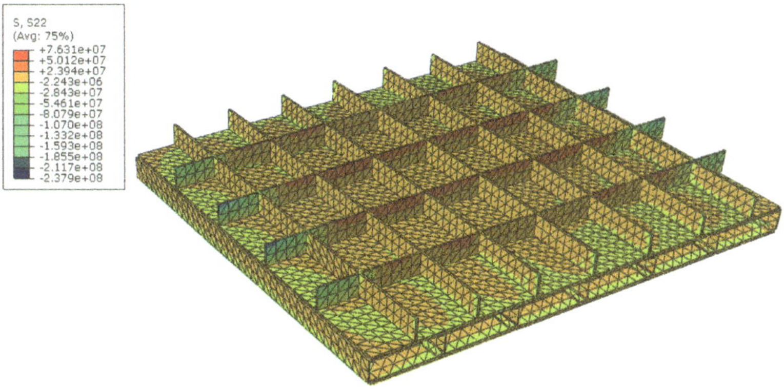

(c) Distribution of σ_y

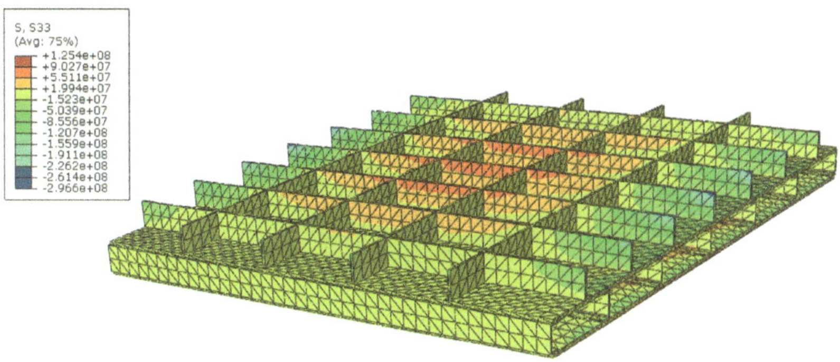

(d) Distribution of σ_z

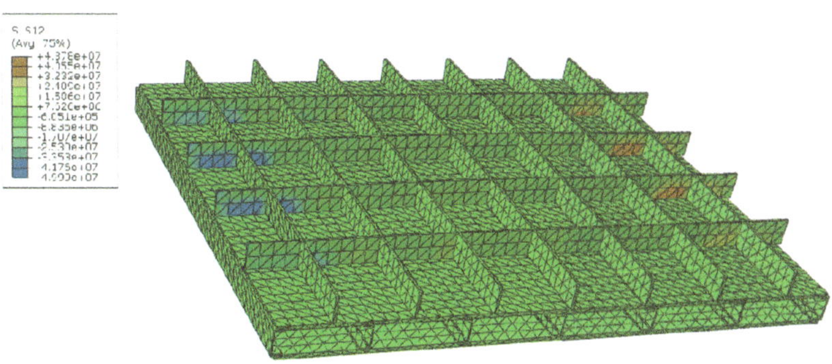

(e) Distribution of τ_{xy}

Fig. 2.20 (continued)

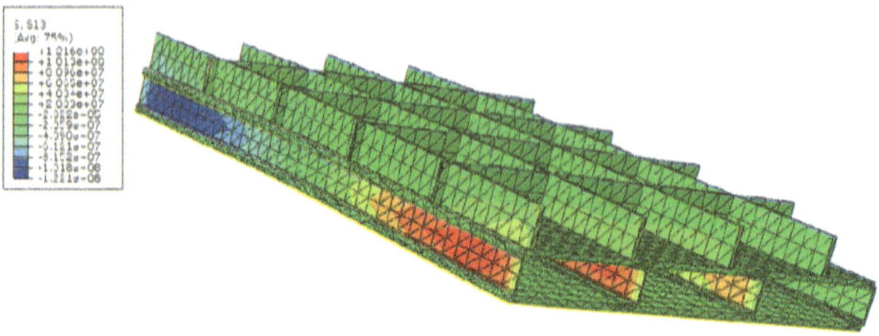

(f) Distribution of τ_{xz}

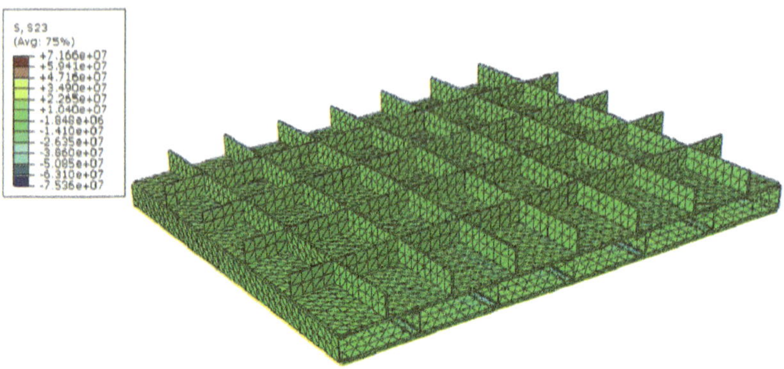

(g) Distribution of τ_{yz}

Fig. 2.20 (continued)

$$\sigma = \frac{Q'}{l_w t} = \frac{1427.0\ \text{kN}}{8.25\ \text{m} \times 0.012\ \text{m}} = 14.4\ \text{N/mm}^2 \leq \text{f}_\text{t}^\text{w} = 185\ \text{N/mm}^2$$

where l_w represents the weld length, which is 8.25 m; t represents the weld thickness, which is 12 mm; f_t^w represents the tensile strength of the weld, which is 185 N/mm^2.

So the strength of the weld meets the requirements.

2.3.3.2 Construction of the Emergency Door

When approaching the excavation time, the emergency door should be installed at the excavation side of the by-pass. And then the pressurization and sealing test are carried out. Before the test, fill the emergency door with water and then pressurize

Fig. 2.21 The emergency door

with the air compressor, whose blowing rate is not less than 6 m^3/min. When the air compressor is still running, the pressure must keep 0.3 MPa. The picture of the emergency door is shown in Fig. 2.21.

2.3.4 Attentions for Freezing Construction

Because the tunnel segments are directly exposed to the air, the heat dissipation will result in the poor freezing effect. Therefore, it is necessary to lay insulation boards and freezing pipes on the outer surface of the segments.

When the freezing wall is closed, the frost heave pressure inside the freezing wall can't be released, which will do harm to the tunnel segments. Therefore, two pressure released vents should be set inside the freezing wall at both ends of the by-pass to release the frost-heave force.

The supporting frames should be set up in the tunnel near the by-pass to limit the impact of frost heave before the closure of the freezing wall.

During the excavation period, the temperature monitoring of the excavation face and the convergence of the freezing wall should be strengthened.

The frost heave and thaw settlement will result in the deformation of the surrounding soil. Therefore, the monitoring of tunnel, buildings on the ground surface and surrounding pipelines should be strengthened during the construction period.

The grouting pipes are pre-buried in the lining to decrease the settlement, as shown in Fig. 2.22.

Fig. 2.22 The pre-buried grouting pipes

2.4 Monitoring of the Freezing Engineering

In the freezing construction of the by-pass, real-time monitoring of the soil temperature and brine temperature is beneficial for us to know the development of the freezing wall and adjust the parameters of brine supply system. In addition, real-time monitoring of the frost heave and thaw settlement of frozen soil is beneficial to control the deformation of the tunnel surrounding the by-pass and the settlement of the buildings on the ground surface. Therefore, in order to control the construction progress in real time and ensure the excavation safety, it is necessary to monitor the development of the freezing wall in real time.

2.4.1 Construction Monitoring of Freezing Holes and Freezing Pipes

(1) Location of the freezing holes
Before construction, the base points and base lines should be set up. Layout the freezing holes accurately according to the construction drawings. The deviation of the freezing holes should not be larger than 100 mm, and the spacing of the freezing holes shouldn't be larger than 150 mm. The actual drilling depth is generally 0.3 m larger than the design depth. When there are steel bars in the structure, the hole position should be adjusted properly to avoid the steel bars.
(2) Inclination survey of the freezing holes
During the construction of horizontal freezing holes, the front end of the drilling tool tends to droop because of its dead weight. In addition, its clockwise rotation will produce dextral force. Both of the two factors result in the opening deflection. The method of lamplight inclinometer is the most advanced technique for the inclination survey of the horizontal freezing holes now, which is suitable for the freezing holes with the drilling depth of no more than 20 m.
(3) Leakage test under high pressure
Check the sealing property of the freezing pipes by injecting water and pressurizing with pressure test pump. The test pressure is 2 times the sum of the pressure difference between the brine column in the freezing pipes and water column and the working pressure of the brine pump. It is required that the pressure drop in 30 min should not exceed 0.05 MPa and then in the next 15 min the pressure keeps unchanged.

2.4.2 Monitoring of the Frost Heave Pressure

The frost heave pressure is an additional stress generated by the volume expansion of the frozen soil, which is a temporary load. But its influence on the surrounding structures is much greater than that of the water and earth pressure.

(1) Monitoring of the pressure relief vents
During the active freezing period, the function of the pressure relief vent is to release the frost heave pressure and judge the closure of the freezing wall. Water migration occurs during the freezing period. And the frost heave pressure can't be released after the closure of the freezing wall. It will increase the water and earth pressure inside the freezing wall. The pressure change of the pressure relief vent will lag behind the closure of the freezing wall for 1–2 days. There are generally 2 pressure relief vents at each end of the by-pass. The change of the frost heave pressure is monitored with a piezometer. When the pressure of the pressure relief vent rises more than 0.2 MPa, it's time to release the water pressure. If the water flows continuously, the valve should be closed immediately and observe it continuously. At the early stage of the freezing period, the pressure relief vents are observed once a day and 2 or 3 times a day when the freezing wall is going to closing.

(2) Monitoring of the frost-heave force
With the increase of the freezing time, the volume of the frozen soil increases continuously. When the frozen soil is restricted by the structure, it will produce a large frost-heave force. Therefore, vibrating steel wire pressure transducers are usually pre-buried at the interface of frozen soil and structure to monitor the frost-heave force.

2.4.3 Monitoring of the Temperature Field

The monitoring of the temperature field includes the monitoring of the brine temperature, the temperature of the temperature holes, the interface temperature of the freezing wall and the segments and the temperature of the prospect holes.

(1) Monitoring of the brine temperature
The brine temperature is monitored with the temperature sensors pre-buried in the freezing system. Monitor once a day.

(2) Temperature monitoring of the temperature holes
The development of the temperature field can be known by monitoring the temperature of the temperature holes. According to the temperature data, the closure of the freezing wall, the thickness and the average temperature of the freezing wall can be correctly judged.

There are 13 temperature holes in total, 5 in one tunnel and 8 in the other, as shown in Figs. 2.23 and 2.24. The characteristic parameters and the number of measuring

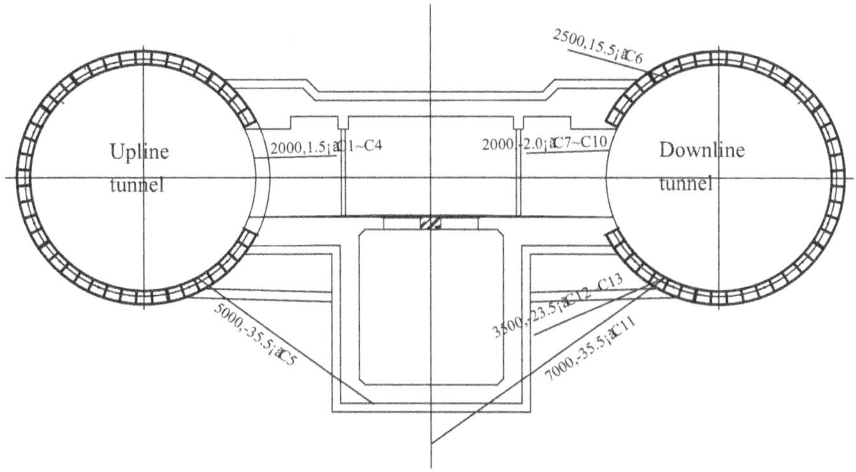

Fig. 2.23 Profile of the temperature hole arrangement

points in each temperature hole are summarized in Table 2.10. The spacing of the measuring points is 500 mm, as shown in Fig. 2.25. Monitor once a day.

(1) Monitoring of the interface temperature of the freezing wall and the segments
When measuring the interface temperature of the freezing wall and the segments, one measuring point should be pre-buried at about 3–5 cm on both sides of the interface. The interface temperature should be determined by interpolation method.

(2) Temperature monitoring of the prospect holes
Before excavation, we need to check the freezing condition at the interface of the freezing wall and the segments by setting prospect holes in the poor freezing area according to the temperature data. The depth of the prospect hole is between 30 and 50 cm. The temperature of the prospect holes shouldn't be higher than −5 °C.

2.4.4 Monitoring of the Deformation

(1) Deformation of the frost heave and thaw settlement on the ground surface
During the artificial freezing construction, there are many factors that will result in the deformation of the ground surface, such as pore-creating construction, frost heave and thaw settlement, excavation and unloading effect, backfill grouting and creep of the freezing wall. The freezing construction will change the displacement field and the stress field. So the monitoring of the deformation should be carried out with high precision leveling and invar tape to avoid large

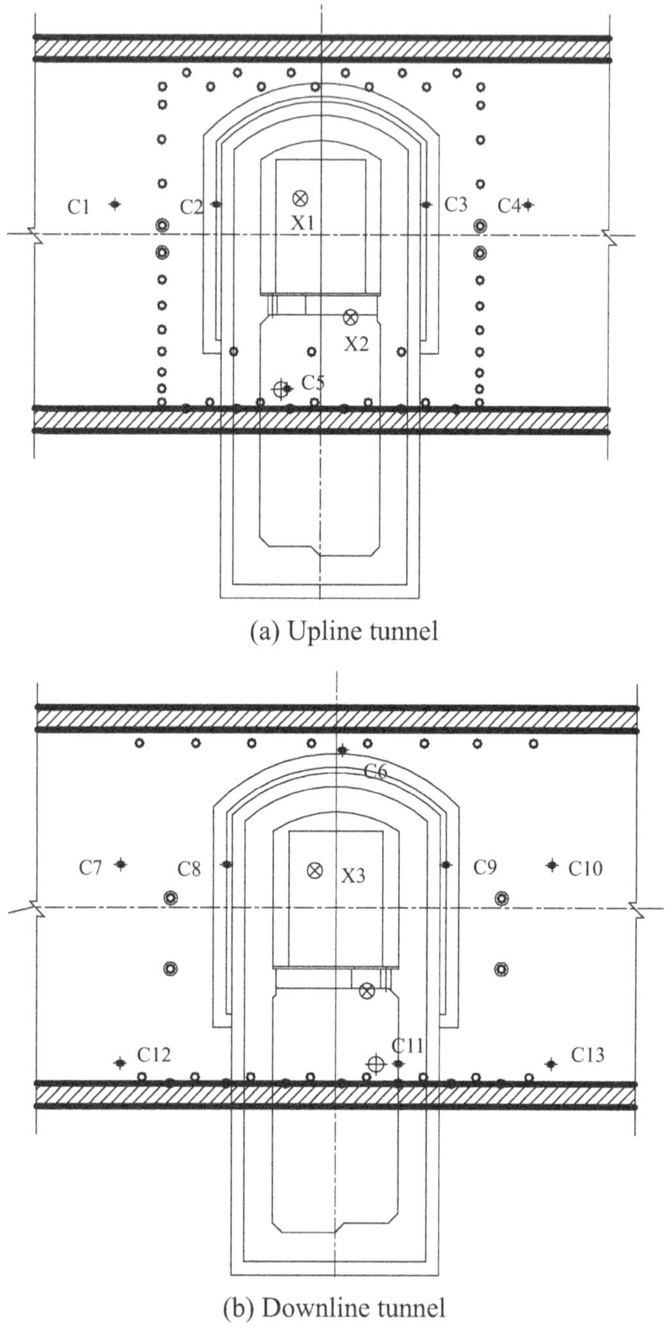

Fig. 2.24 Elevation of the temperature hole arrangement

Table 2.10 The characteristic parameters and the number of measuring points in each temperature hole

Number	Diameter (mm)	Length (m)	Inclination (°)	Number of measuring points
C1–C4	Φ32	2.0	1.5	4
C5	Φ89	5.0	−35.5	10
C6	Φ32	2.5	15.5	5
C7–C10	Φ32	2.0	−2.0	4
C11	Φ89	7.0	−35.5	14
C12–C13	Φ32	3.5	−23.5	7

Note The dip angle is the angle between the temperature hole and the horizontal plane. The elevation angle is positive and the depression angle is negative

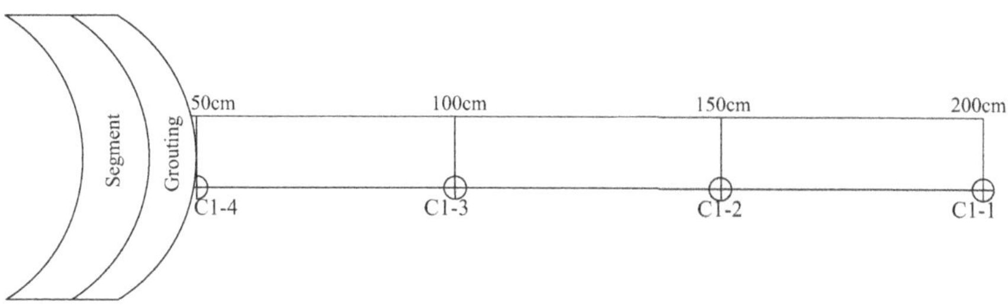

Fig. 2.25 Diagram of the temperature measuring point arrangement

impact on the surrounding environment. The measuring points are arranged at the top and both sides of the by-pass and the spacing is 10 m.

(2) Deformation of surrounding buildings and pipelines

The deformation monitoring points of the buildings should be pre-buried in the places that can reflect the deformation characteristics of the buildings and places with obvious deformation. The deformation of the buildings includes the vertical displacement and cracks.

For buildings subjected to the influence of frost heave, the vertical displacement is monitored by placing measuring points on the external walls or support pillars. The spacing of the measuring points on the external walls should be 10–15 m. Or there is one measuring point for every 2 bearing columns. For buildings subjected to little influence of frost heave, the spacing of the measuring points on the external walls should be 15–30 m. Or there is one measuring point for every 2–3 bearing columns.

The method of crack monitoring is setting at least 2 monitoring points for each crack, one set at the widest place of the crack and the other set at the end of the crack.

The measuring method of the pipeline deformation is fixing monitoring points on the pipeline and extending the mark to the ground surface. The measuring points should be pre-buried in the places with obvious deformation. The spacing of the measuring points should be 5–15 m for main influenced area or 15–30 m for less influenced area. The deformation of the pipeline includes vertical displacement and differential settlement.

(3) Horizontal convergence and settlement of the tunnel
The monitoring datum points of the horizontal displacement and vertical displacement are arranged in the stable zone 50 m away from the by-pass.

a. Horizontal convergence
During the freezing construction, the frost heave will have impact on the existing tunnel. Therefore, the monitoring points should be set up on the segments and marked with red paint. The convergence gauge is used to monitor the convergence of the existing tunnel, which shouldn't be more than 10 mm.
b. Settlement
The settlement monitoring points are set up at the bottom segments and the spacing is 2 m. The settlement of each point is monitored with the precision level or total station. The settlement of the tunnel shouldn't exceed 10 mm, and the settlement per day shouldn't exceed 3 mm.
c. Monitoring of the holding power of the pre-stressed support
The frost heave and thaw settlement will do harm to the tunnel. In order to reduce the influence on the tunnel deformation, two pre-stressed supports shall be installed on each side of the by-pass. The pre-stressed supports are installed at the first tunnel segment joints near the by-pass symmetrically. There are 7 or 8 fulcrums for each pre-stressed support. The pre-stress is provided by the screw jacks. The pressure cell is placed between the jack and the tunnel segment. In order to prevent the stress concentration, the steel plates are mounted on the contact surface of the pressure cell and the jack and the contact surface of the pressure cell and the tunnel segment. Use the frequency recorder to obtain the data from the pressure cell and convert it to pressure.

(4) Horizontal convergence and settlement of the by-pass
The arrangement of the monitoring points and measuring method for the deformation in the by-pass are the same as those in the tunnel.
Fig. 2.26 shows the monitoring points arrangement of the ground deformation above the by-pass. The number of monitoring points is summarized in Table 2.11. The allowable values of the deformation are summarized in Table 2.12. Figure 2.27 shows the monitoring points arrangement of the deformation in the tunnel.
(5) The monitoring frequency is summarized in Table 2.13.

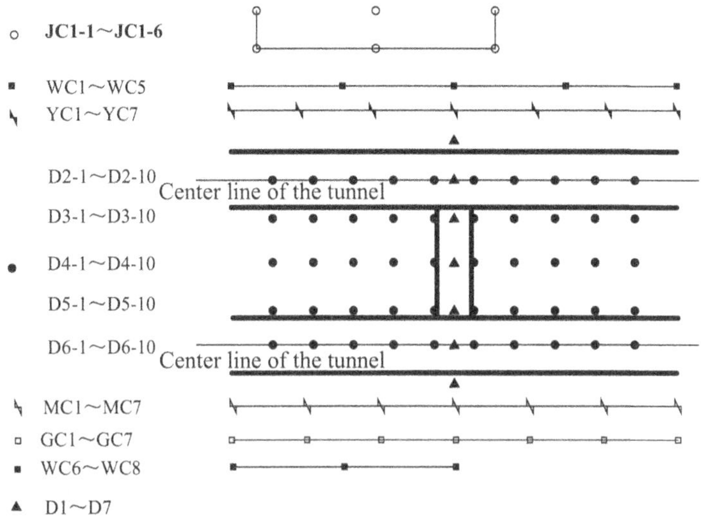

Note: ○ represents the settlement monitoring points of the surrounding buildings; ■ represents the settlement monitoring points of the sewage conduits; ↯ represents the settlement monitoring points of the storm sewers; ● represents the settlement monitoring points of the tunnel; ↯ represents the settlement monitoring points of the gas conduits; □ represents the settlement monitoring points of the water supply conduits and ▲ represents the settlement monitoring points of the by-pass.

Fig. 2.26 The plan of the monitoring points arrangement of the ground deformation above the by-pass

Table 2.11 The number of monitoring points

Type	Number	Serial number
Ground surface	57	D2-i–D6-i(i:1–10), D1–D7
Surrounding buildings	6	JC1–1–JC1–6
Water supply conduits	7	GC1–GC7
Sewage conduits	8	WC1–WC8
Gas conduits	7	MC1–MC7
Storm sewers	7	YC1–YC7

2.4.5 Monitoring of the Freezing Equipment

The monitoring of the freezing equipment mainly includes the cooling water temperature, suction and discharge temperature of the freezer, working pressure of

Table 2.12 The allowable values of the deformation

Type	Monitoring targets	Judgment basis	Allowable values (absolute values)		
			Accumulated value (mm)		Rate of variation (mm/d)
			Early-warning value	Alarm value	
Surrounding soil	Settlement of the roads and ground surface	The absolute variation of the elevation	70%	−20, + 10	±3
Surrounding pipelines	Settlement of the pipelines			±10	±1
Surrounding buildings	Buildings			±15	±1

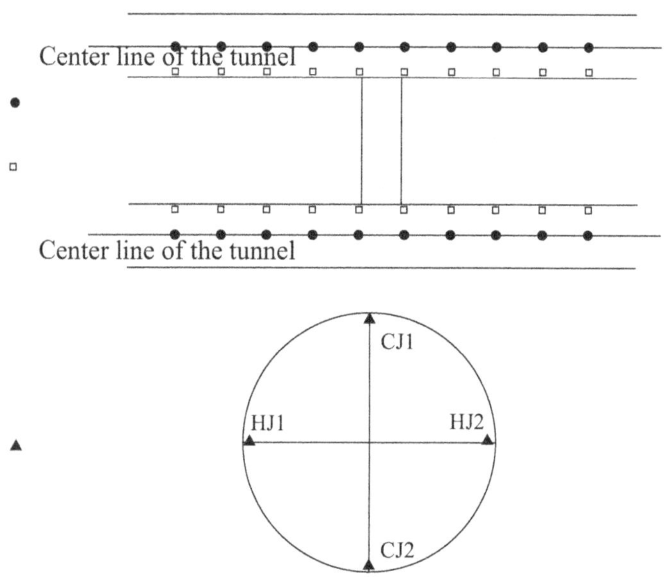

Note: ● represents the settlement monitoring points; □ represents the displacement monitoring points and ▲ represents the convergence monitoring points.

Fig. 2.27 The plan of the monitoring points arrangement of the deformation in the tunnel

brine pump, suction and discharge pressure of the freezer, condensing pressure of refrigeration system and vaporization pressure of refrigeration system, etc.

The cooling water temperature and the suction and discharge temperature of the freezer are measured with the infrared temperature or precise mercury temperature. Monitor once a day. The working pressure of the refrigeration system and brine pump are measured with the pressure gauge for ammonia or general pressure gauge. The

Table 2.13 The monitoring frequency

Monitoring targets	Drilling period	Freezing period	Excavation period	Thawing period
Vertical displacement monitoring of integrated pipelines	Once a day	Once every other day	Once a day	Once per 2–5 days for the first 3 months; Once per 5–10 days for the 4th and 5th month; Once per 10–15 days for the 6th month
Vertical displacement monitoring of surrounding buildings	Once a day	Once every other day	Once a day	
Vertical displacement monitoring of the ground surface	Once a day	Once every other day	Once a day	
Vertical displacement monitoring of the tunnel	Once every other day	Once every other day	Once a day	
Convergence monitoring	Once every other day	Once every other day	Once a day	

flowmeter is used to measure the total brine flow. And the brine flow of each freezing pipe shouldn't be smaller than 5 m^3/h. The liquid level monitoring device is installed in the brine tank to prevent the loss of brine.

2.5 In-Site Monitoring of the Temperature Field

In order to understand the development of the freezing wall and make sure the thickness and the average temperature of the freezing wall meet the design requirements, it is necessary to monitor and analyze the temperature field during the freezing construction. In addition, the frozen soil has the characteristic of frost heave and thaw settlement, so it is necessary to monitor and analyze the deformation of the ground surface.

2.5.1 Analysis of the Brine Temperature

Figure 2.28 shows the brine temperature curve during the active freezing period. The variation process of the brine temperature can be divided into three stages.

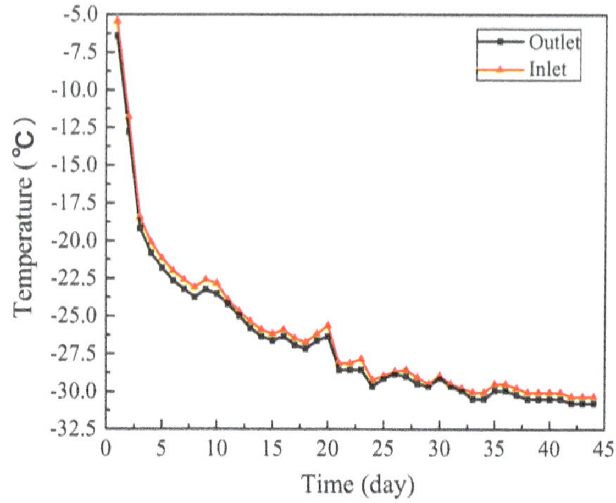

Fig. 2.28 The cooling curve of the brine

(1) Rapid cooling stage
At the beginning of the freezing process, the brine temperature decreases rapidly because of the high brine temperature and the high cooling efficiency. At day 4, the brine temperature is under −18 °C, which meets the design requirement of decreasing to −18 °C at day 7. The brine difference between outlet and inlet is 0.8–1.1 °C.

(2) Slow cooling stage
As the brine temperature decreases, the cooling efficiency of the refrigerator is reduced and the cooling capacity is reduced. At the same time, as the frozen cylinder around the freezing pipe expands continuously, it hinders the release of the cooling capacity in the freezing pipe. At day 11, the brine temperature is under −24 °C, which meets the design requirement of decreasing to −24 °C at day 15. The brine difference between outlet and inlet is 0.5–0.7 °C.

(3) Stable stage
At this stage, the frozen cylinders have been enclosed so that the expansion speed of the frozen cylinder decreases gradually and the thermal resistance of the frozen cylinder increases slowly. Therefore, the brine temperature is stable because the cooling capacity of the freezer is balanced with the release of the cooling capacity of the freezing pipes. The brine temperature decreases to −30 °C gradually and the brine difference between outlet and inlet is 0.5 °C.

From Fig. 2.28, we can see that the brine difference between outlet and inlet is under 1.1 °C, which means the cooling capacity of the freezer is abundant. At the early stage, the temperature difference between the brine and soil is large and the heat exchange is fast, so the temperature difference between the outlet and inlet is large. As the soil temperature decreases gradually, the heat exchange between the soil and brine decreases gradually and the temperature difference between outlet and inlet tends to be stable. At day 19, the brine temperature goes up because of the machine halt.

2.5.2 *Analysis of the Frozen Soil Temperature*

Figure 2.29 shows the temperature variation of temperature hole C1 and C2. At the beginning, the descent speed of the soil temperature is large. Then it becomes small gradually and finally tends to be stable. This is consistent with the decrease of the brine difference between outlet and inlet. The soil temperature is approximately logarithmic with the freezing time. At the beginning of the freezing period, the temperature of the soil near the segment is lower than the temperature of the inner soil because it is winter and the soil near the segment is influenced by the air temperature. However, with the increase of the freezing time, the frozen cylinders will be enclosed

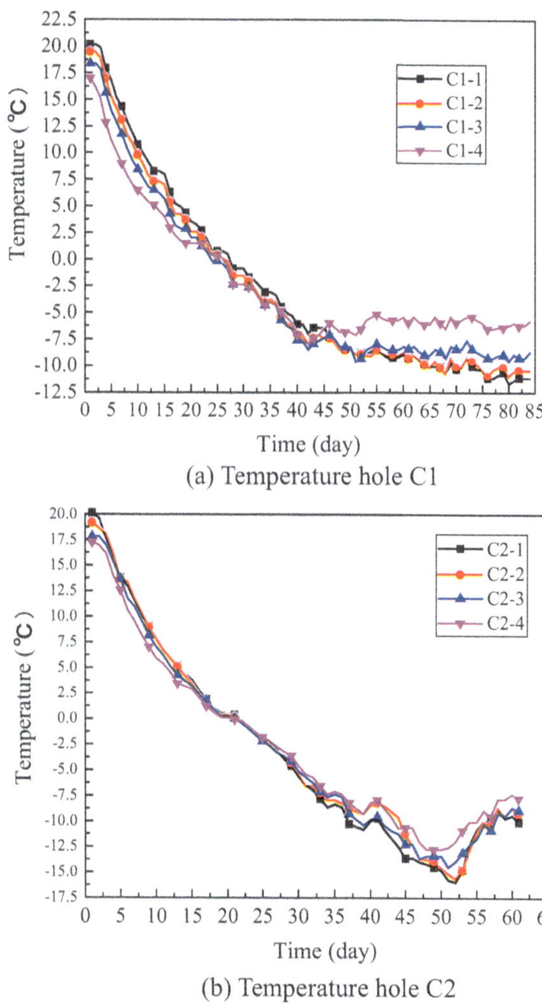

Fig. 2.29 Temperature variation curve of different temperature holes

so that the temperature of the soil near the segment tends to be higher than the temperature of the inner soil. At day 50, the soil excavation is carried out. From Fig. 6.25b, we can see that the soil temperature is going up. Because the temperature hole C2 is on the inside of the freezing wall. And when the inner soil is excavated, the soil around the temperature hole C2 is exposed to the air. The heat exchange is intense, which results in the rise of the soil temperature. Therefore, it is necessary to strengthen the heat preservation of the excavation face. The temperature hole C2 is removed at day 61. From day 18, the temperature drop is nearly equal to zero because the water in the soil is in the phase transformation stage, which lasts about 4 days. This indicates that the frozen cylinders are enclosing.

Figure 2.30 shows the temperature curve of each measuring point at the depth of 150 cm varying with time. Those measuring points belong to the eight temperature holes at the opposite of the freezing station. The distances between the temperature holes and the freezing pipes are shown in Table 2.14. The temperature holes C7, C10, C12 and C13 are on the outside of the freezing wall and the others are on the inside of the freezing wall.

From Fig. 2.30, the temperature drop of the temperature hole C6 is the fastest and keeps the lowest, because the temperature hole C6 is the closest to the freezing pipe. The temperature drops of the temperature holes C7 and C10 are the slowest and their temperatures are always the highest, because the temperature holes C7 and C10 are on the outside of the freezing wall and adjacent to the unfrozen soil. The

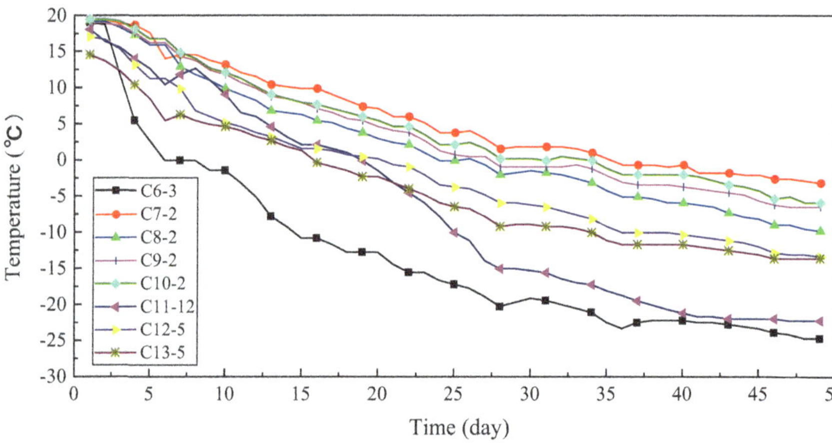

Fig. 2.30 The temperature curve of each measuring point at the depth of 150 cm varies with time for the temperature holes at the opposite of the freezing station

Table 2.14 The distances between the temperature holes and the freezing pipes

Temperature hole	C6	C7	C8	C9	C10	C11	C12	C13
Distance (mm)	43	110	106	108	110	70	70	70

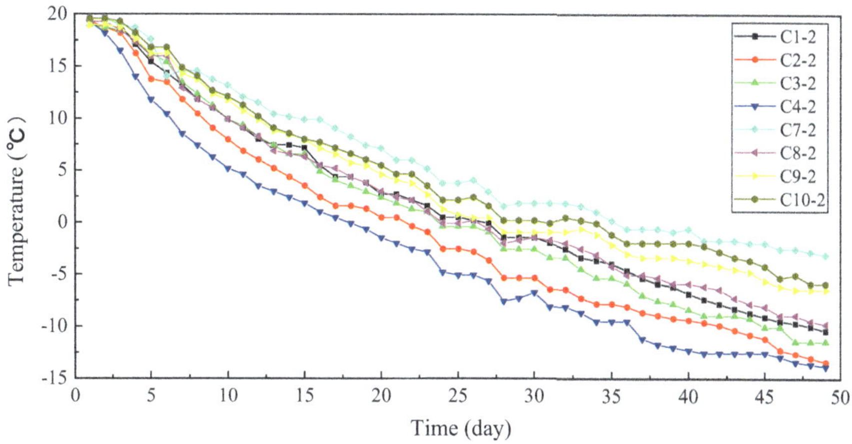

Fig. 2.31 The temperature curve of each measuring point at the same depth varies with time for temperature holes on both sides of the by-pass

temperature holes C7 and C10 are greatly influenced by the unfrozen soil so that the cooling capacity cannot be gathered. Temperature holes C7, C10, C12 and C13 are all on the outside of the freezing wall. However, compared with the temperature holes C7 and C10, the temperatures of temperature holes C12 and C13 drop faster and their temperatures are always lower. This is because the temperature holes C12 and C13 are closer to the freezing pipes.

Figure 2.31 shows the temperature curve of each measuring point at the same depth varying with time for temperature holes on both sides of the by-pass. The temperature of each temperature hole on the side of the freezing station drops faster and their final temperatures are lower. Because the brine circulation pipeline on the side of the freezing station is shorter, and the resistance is smaller. Under the same brine pump pressure, the brine flow rate is faster and the heat exchange is more frequent.

2.5.3 Calculation of Thickness and Average Temperature of Freezing Wall

The main function of the freezing wall is bearing the water pressure and earth pressure. The bearing capacity of the freezing wall is determined by the thickness of the freezing wall and the distribution of the temperature field. Therefore, the thickness and average temperature must meet the requirements before excavation.

Because the axial heat conduction ratio of the freezing pipe is much smaller than the radial heat conduction ratio, the temperature field of the single pipe can be

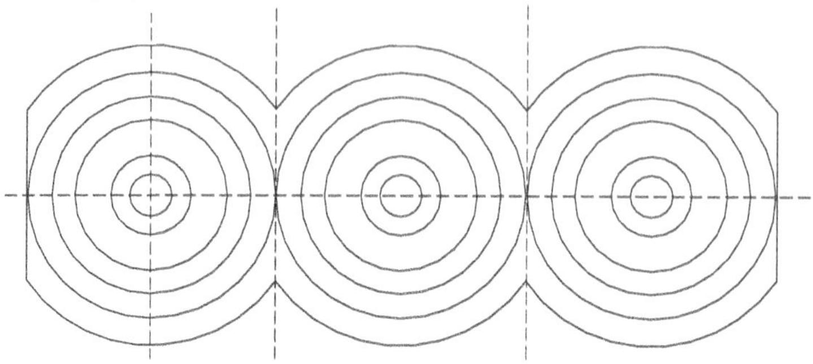

Fig. 2.32 The change of the isotherm during the freezing process

simplified as a plane heat conduction problem. The change of the isotherm during the freezing process is shown in Fig. 2.32.

The average temperature of the freezing wall is the average of the temperature distribution on a certain cross section or the average temperature of the whole freezing wall. When the freezing wall is homogeneous, it is the average temperature of the whole freezing wall. When a specific cross section is paid attention to, it is the average of the temperature distribution on the cross section. When the waterproof performance between the freezing wall and the structure is paid attention to, it is the average temperature of the freezing wall between the freezing wall and the structure.

In order to meet different working conditions, the arrangement of the freezing pipes is also different. The common arrangement is shown in Fig. 2.33. Generally speaking, the weak section of the freezing wall is the side wall because of the arrangement of the single-row freezing pipes. Therefore, the following formulas for the thickness and the average temperature of the single-row freezing pipes are mainly studied.

2.5.3.1 The Common Method Used in the Engineering

The radius of the frozen cylinder is

$$r_2 = \exp\left(\frac{t_1 \ln r - t \ln r_1}{t_1 - t}\right) \tag{2.10}$$

where r_2 is the radius of the frozen cylinder (m); t_1 is the inlet brine temperature, which is −30.5 °C; r is the distance between the temperature hole and the freezing pipe, which is 1.10 m; r_1 is the inside radius of the freezing pipe, which is 0.0365 m and t is the temperature of the measuring point, which is −3.1 °C.

Considering the temperature of the measuring point and the distance between the temperature hole and the freezing pipe, the temperature of the measuring point C7-1

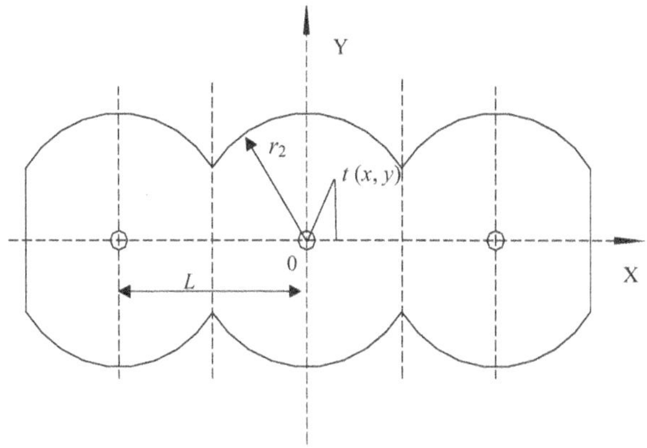

(a) The schematic diagram of the temperature field for single-row freezing pipes

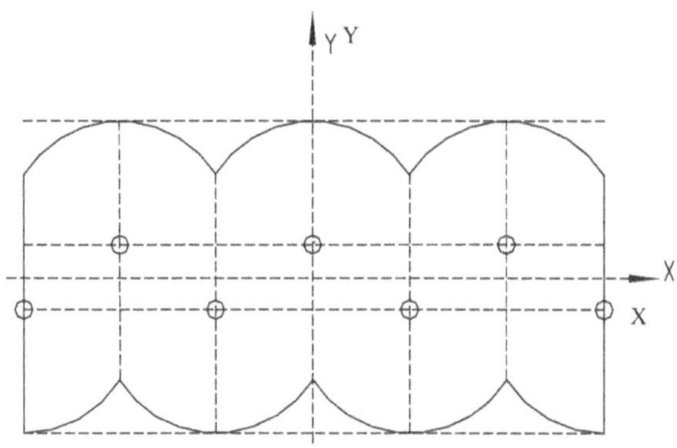

(b) The schematic diagram of the temperature field for double-row interlaced freezing pipes

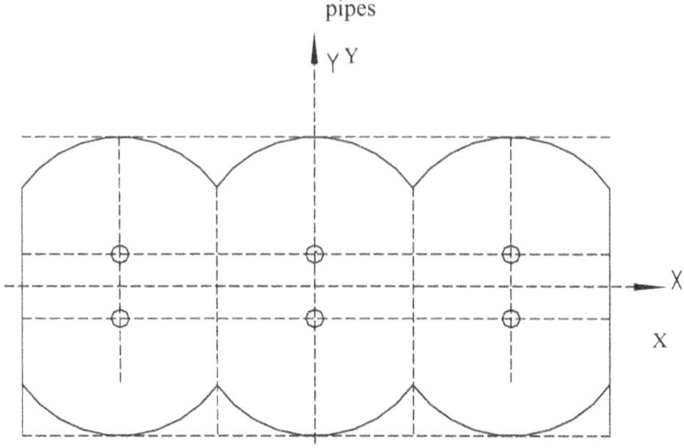

(c) The schematic diagram of the temperature field for double-row freezing pipes

Fig. 2.33 The common arrangement of the freezing pipes

is higher. The distance is 1.10 m and the temperature is −3.1 °C. Substituting the distance and the temperature into Eq. (6.10), we can obtain that the minimum radius of frozen cylinder is 1.62 m.

According to the Pythagorean theorem, the thickness of the freezing wall is

$$E = 2\sqrt{r_2^2 - \left(\frac{L}{2}\right)^2} \tag{2.11}$$

where E is the thickness of the freezing wall (m) and L is the spacing of the freezing holes, which is 0.90 m.

Therefore, the minimum thickness of the freezing wall is 3.11 m.

The formula for average temperature of the freezing wall is

$$t_{oc} = t_1 \times \left(1.135 - 0.352\sqrt{L} - 0.875\frac{1}{\sqrt[3]{E}} + 0.266\sqrt{\frac{L}{E}}\right) - 0.466 \tag{2.12}$$

$$t_c = t_{oc} + 0.25t \tag{2.13}$$

where t_{oc} is the average temperature of the freezing wall when the freezing temperature of soil is zero (°C) and t_c is the average temperature within the effective thickness of the freezing wall when the surface temperature of the side wall after excavation is below zero (°C).

Therefore, the average temperature of the freezing wall is −11.8 °C.

2.5.3.2 Development Speed of the Frozen Cylinder

At day 37, the temperature of the measuring point C7-1 reaches 0 °C for the first time. The distance between the temperature hole and the freezing pipe is 1.10 m. So the development speed of the frozen cylinder is 0.0297 m/d. The excavation is carried out at day 47, and at this moment the radius of the frozen cylinder is 1.40 m. According to Eq. (2.11), the thickness of the freezing wall is 2.65 m. According to Eqs. (2.12) and (2.13), the average temperature is −11.1 °C.

2.5.3.3 Temperature Gradient Method

As is shown in Fig. 2.34, the distance between the temperature hole and the freezing hole is 1.10 m. The temperature of the temperature hole is −3.1 °C and the temperature of the freezing hole is −30.5 °C. Therefore, according to the similar triangle method, the position of 0 °C is 1.22 m away from the freezing hole. So the radius of the frozen cylinder is 1.22 m. According to Eq. (2.11), the thickness of the freezing wall is 2.27 m. According to Eq. (2.12) and Eq. (2.13), the average temperature is −10.5 °C.

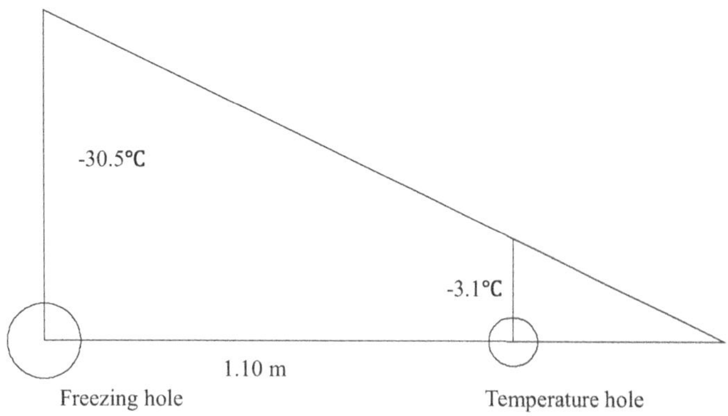

Fig. 2.34 Temperature gradient

2.5.3.4 The Revised Bakholdin Formula

Bakholdin believes that the temperature field distribution of the single-row freezing pipes can be assumed to be a steady two-dimensional temperature field. The wavy surface of the initial freezing wall will soon be filled so that the surface of the freezing wall can be assumed to be flat. Based on the heterogeneous similarity of the thermodynamics and hydraulics, the analytic solution of the temperature field of the single-row freezing pipes is obtained. However, the actual freezing temperature of soil is lower than 0 °C because of some soluble salts in the soil and the influence of earth pressure. Therefore, Huang (2008) proposed the revised Bakholdin formula. The temperature field of the single-row freezing pipes is

$$t_{xy} = \frac{t_1 - t_0}{\ln \frac{L}{2\pi r_1} + \frac{\pi r_2}{L}} \cdot \left[\frac{\pi r_2}{L} - \frac{1}{2} \ln 2 \left(ch \frac{2\pi y}{L} \right) - \cos \frac{2\pi x}{L} \right] + t_0 \tag{2.14}$$

$$E = 2r_2 \tag{2.15}$$

where r_2 is 1.62 m and t_0 is the freezing temperature of the soil, which is −0.92 °C.

Huang (2008) obtained the formula of the average temperature at the interface of the freezing wall for single-row freezing pipes through lots of experiments.

$$t_{\text{cpi}} = \frac{1}{2}(t_{\text{k}} + t_0) = \frac{1}{2}\left(t_{\text{x}=\frac{L}{2},\text{y}=0} + t_0\right) \tag{2.16}$$

where t_{cpi} is the average temperature at the interface of the freezing wall (°C) and t_{k} is the temperature at the intersection of the axial plane and the interface of the freezing wall (°C).

According to Eqs. (2.11)–(2.13), the thickness of the freezing wall is 3.24 m and the average temperature is −12.1 °C.

The designed thickness of the freezing wall is 1.8 m and the designed average temperature is −10 °C. The calculation results show that the freezing wall meets the design requirements. In addition, the development of the freezing wall can be judged by observing the frosting range and contour on the segments near the by-pass.

According to the measured data, the calculated thickness of the freezing wall is 26–80% larger than the designed thickness, while the average temperature of the freezing wall is only 11–21% larger than the designed thickness. This result indicates that compared with the thickness of the freezing wall, the average temperature is more difficult to meet the design requirements. Therefore, the average temperature of freezing wall is the decisive factor of whether the freezing wall can meet the design requirements. And the average temperature of freezing wall is an important factor that affects the strength and stiffness of freezing wall. It is noteworthy that the average temperature of the freezing wall is related to the initial freezing temperature of the soil, the brine temperature and the spacing of the freezing pipes. Therefore, reducing the brine temperature or the spacing of the freezing pipes should be considered firstly to improve the strength and stiffness of the freezing wall.

2.5.3.5 Analysis of the Revised Bakholdin Formula

When there is only one row freezing pipes, the range of r_2/L in actual engineering is between 0.5 and 2. Four cases of $r_2/L = 0.5$, 1, 1.5 and 2 are used to analyze the change of average temperature at the interface of the freezing wall, as shown in Fig. 2.35. The average temperature at the interface of freezing wall is

$$t_{\text{cpi}} = \frac{1}{2}(t_k + t_0) = \frac{1}{2}\left(t_{x=\frac{L}{2},y=0} + t_0\right)$$

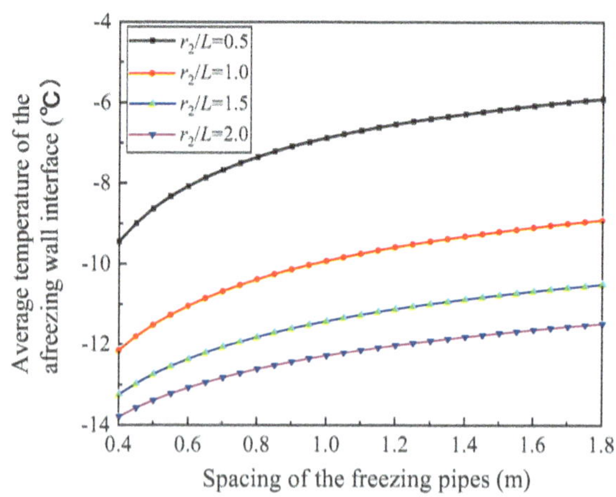

Fig. 2.35 Under the conditions of different r_2/L, the average temperature of the freezing wall varies with the spacing of the freezing pipes

$$
\begin{aligned}
&= \frac{1}{2} \frac{-30.5 + 0.92}{\ln \frac{L}{2\pi \times 0.0365} + \frac{\pi r_2}{L}} \times \left[\frac{\pi r_2}{L} - \frac{1}{2} \ln 2 \right] - 0.92 \\
&= \frac{-14.79}{\ln \frac{L}{0.23} + \frac{\pi r_2}{L}} \times \left[\frac{\pi r_2}{L} - \frac{1}{2} \ln 2 \right] - 0.92
\end{aligned} \tag{2.17}
$$

From Fig. 2.35, we can obtain that: (1) when r_2/L is constant, the average temperature rises with the increase of the spacing of the freezing pipes. So the spacing of the freezing pipes can't be too large; (2) when the spacing of the freezing pipes is constant, the distance between curves become smaller with the increase of r_2/L, which means that with the increase of the radius of the freezing cylinders, the temperature drop decreases gradually.

2.5.4 *Vertical Displacement of Ground Surface During Active Freezing Period*

2.5.4.1 Vertical Displacement of Ground Surface in Axial Direction of by-Pass

Figure 2.36 illustrates the vertical displacement of the ground surface in the axial direction of the by-pass. At day 1, all the measuring points are subsiding because a small amount of soil was taken out when drilling the freezing holes. With the increase of the freezing time, the heave of each measuring point is on the rise because of the frost heave. But the accumulative heave is less than 10 mm. According to Table 2.12, it is safe. The maximum frost heave is happened at measuring point D4, so is the frost heave growth rate. The minimum frost heave is happened at measuring points

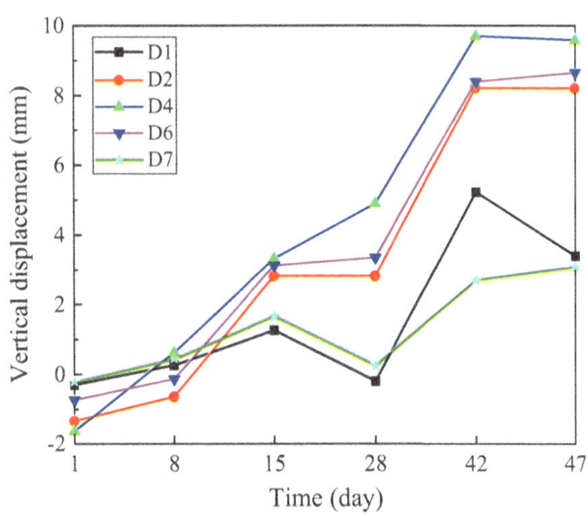

Fig. 2.36 The vertical displacement of the ground surface in the axial direction of the by-pass

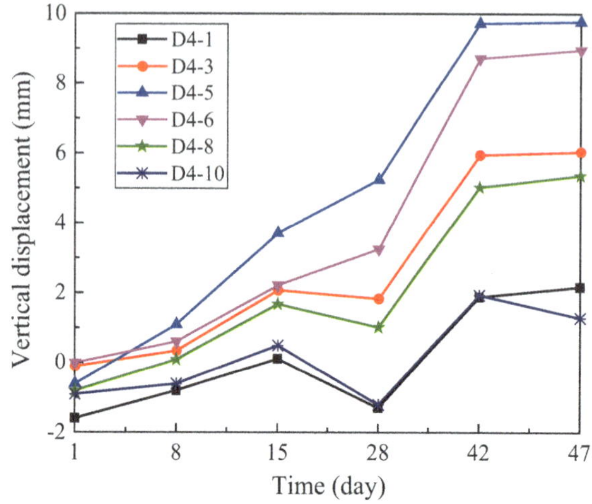

Fig. 2.37 The vertical displacement of the ground surface perpendicular to the axial direction of the by-pass

D1 and D7, so is the frost heave growth rate. Because the measuring point D4 is located on the center point along the axis of the by-pass, while D1 and D7 are at both ends of the axis, as is shown in Fig. 2.26. This means that the closer to the freezing center, the greater the influence of frost heave.

2.5.5 Vertical Displacement of Ground Surface Perpendicular to Axial Direction of by-Pass

Figure 2.37 illustrates the vertical displacement of the ground surface perpendicular to the axial direction of the by-pas. The frost heave of the ground surface is distributed symmetrically and the axis of the by-pass is the symmetric axis. The maximum frost heave is happened at the axis of the by-pass, so is the frost heave growth rate. But it is safe. As is shown in Fig. 2.26, the closer to the axis of the by-pass, the greater the frost heave and its growth rate.

2.5.6 Maximum Deformation and Change Rate of Various Monitoring Projects

During the active freezing period, the accumulative maximum value and maximum change rate of each monitoring project are summarized in Table 2.15. The accumulative maximum frost heave is happened at the axis of the by-pass, so is the frost heave growth rate. The accumulative maximum values of the frost heave happened

Table 2.15 The accumulative maximum value and maximum change rate

Monitoring project	Number	The accumulative maximum value (mm)	Maximum change rate (mm/d)
The axis of the by-pass	D5	9.98	0.11
The cross section of the by-pass	D4–5	9.78	0.11
Water supply conduits	GC5	4.07	0.05
Sewage conduits	WC5	4.02	0.05
Storm sewers	YC3	4.02	0.05
Gas conduits	MC2	5.07	0.06

at the location of the rigid pipelines are small because the rigid pipelines are far away from the by-pass, so are the growth rates.

Therefore, in order to reduce the influence of the frost heave, pressure-relief vents can be set within the freezing wall. At the same time, adjust the brine flow and brine temperature to control the thickness of the freezing wall so that the impact on the tunnel segments, underground pipelines and the surface structures can be reduced.

2.6 Conclusion

In this chapter, the artificial freezing method used in the construction of the by-pass in the silty clay stratum is introduced. The design of freezing parameters, construction technologies, monitoring of the temperature field and deformation are introduced systematically. The results include the following:

(1) The freezing parameters of the by-pass are designed, mainly including the thickness and the average temperature of the freezing wall, the freezing hole, the brine temperature and the flow rate.
(2) The force-method and finite element software are used to design the thickness of the freezing wall. The former result is 3.5 m while the latter demonstrates that the freezing wall with thickness of 1.8 m can meet the requirements, which indicates that the force-method is conservative.
(3) The total number of the freezing holes is 74. The type of the freezing pipe is $\phi 89 \times 8$. The design value of the average temperature of the freezing wall is -10 °C. The lowest brine temperature must not be higher than -28 °C. The brine flow of the single hole shouldn't be smaller than 3 m^3/h.
(4) During the freezing period, the descent speed of the soil temperature and brine temperature is large at the beginning. Then it becomes small gradually and finally tends to be stable. The temperature of the soil inside the freezing wall drops faster and the final temperature is lower, so is the temperature of the soil on the side the freezing station.

(5) The soil at the center of the by-pass is subjected to the largest influence of the frost heave, so is the growth rate of the frost heave. The farther away from the center, the smaller the influence of frost heave, so does the growth rate of frost heave.
(6) According to the measured data, the calculated thickness of the freezing wall is 26–80% larger than the designed thickness, while the average temperature of the freezing wall is only 11–21% larger than the designed thickness. This result indicates that compared with the thickness of the freezing wall, the average temperature is more difficult to meet the design requirements.

References

Huang F. (2008). *Research on calculation model of artificial frozen wall in saline strata* (pp. 56–57). Tongji University.

Jin, W. W., Chen, Y. L., Li, L., & Wang, L. M. (2008). FEM Analysis of 3D temperature field in connection aisle between tunnels constructed using artificial freezing method. *Journal of Shanghai University (Natural Science Edition), 2*(1), 85–90.

Zheng, G. (2011). *Underground engineering*. Beijing: China Architecture & Building Press.

Zhan, Z. X., Cui, Z.D., Yang, P., & Zhang T. (2020). IN situ monitoring of temperature and deformation fields of a tunnel cross passage in Changzhou Metro constructed by AGF. *Arabian Journal of Geosciences, 13*, 310.

Chapter 3
Axial Strain of Silty Clay Before and After Freezing and Thawing

3.1 Introduction

The soft clay characterized by high water content, high compressibility, high sensitivity, low permeability and low bearing capacity has been recognized as one of the most problematic soils and is widely distributed in the southeast coastal areas of China. With the development of urban rail transposition in these areas, the dynamic response of saturated soft clays under the subway vibration loading has been a point of interest particularly in the last decade due to the obvious land subsidence.

To guarantee the safety in the construction of subway tunnel in soft soil areas, the artificial freezing method has been widely used in the by-pass of subway tunnel. The soil was strengthened when the soft clay was artificially frozen, which did a great favor to the construction (Ling et al. 2013). But the physical and mechanical properties of clay soil, such as water content, density, elastic modulus, porosity and soil macrostructures et al., changed significantly when the frozen clay melt, especially the elastic modulus (Chamberlain and Blouin 1978; Eigenbrod 1996; Konrad 1989). Cui et al. (2014) studied the microstructures of silty clay before and after freezing and thawing by quantitatively analyzing SEM images. The distributions of pore orientation angles of soils before and after freezing and thawing were quite uneven and the soils became looser after freezing and thawing. The elastic modulus was reduced remarkably after freezing and thawing, which contributed to the deformation of soils under the vibration loading and resulted in the uneven settlement of ground (Simonsen and Isacsson 2001).

Tang et al. (2008) and Ren et al. (2011) studied the frequencies of subway vibration loading by in-site monitoring. When the subway train comes near the monitoring site, the vibration loading generates and spreads in waves, and the dynamic response of the soil arises. There are the maximum and the minimum frequencies of the soil response around the subway tunnel. The statistic values are 2.4–2.6 Hz in the maximum and 0.4–0.6 Hz in the minimum, respectively. Matsui et al. (1992) carried out cyclic triaxial tests with Senri clay under frequencies between 0.02 Hz and 0.5 Hz. The

Z.-D. Cui et al., *Dynamics of Freezing-Thawing Soil around Subway Shield Tunnels*,
https://doi.org/10.1007/978-981-15-4342-5_3

lower the frequency was, the higher the pore pressure and the axial strain were when the number of vibration is equal. However, the contrary conclusion was obtained by Yasuhara et al. (2003) when the cyclic triaxial tests were conducted with Ariake clay under frequencies between 0.1 Hz and 10 Hz. The influence of number of vibration on soil strength and stiffness degradation was also studied. The limit of cyclic stress ratio, which was defined as the largest cyclic stress ratio in the process of soil failure, was proposed by Larew and Leonards (1962) for the saturated soft clay, which was proved by Sangrey et al. (1978) through dynamic triaxial tests. The limit of cyclic stress ratio was an important factor to analyze the dynamic response of the soil.

A series of closed hysteresis loops, which gradually moved to the right with the increase of cyclic stress ratio and the number of cycles, were showed in the dynamic stress-strain curves of the cyclic tests (Kokusho 1980). The cumulative plastic strain increased in the process, which was consistent with the results obtained by Yasuhara et al. (1982); Matasovic and Vucetic (1995). The initial static deviator stress was also an important factor influencing stress-strain response of the soil under the dynamic loading. The shear strength of soil was reduced with the increase of the initial static deviator stress (Seed and Chan 1966). However, Lefebvre and Pfendler (1996) found that the shear strength of soil was reduced on the one hand with the applied of the initial static deviator stress, the total strength was improved by 30% approximately on the other hand.

Of course there were still other factors that had a great influence on the dynamic characteristics of clay soil, such as the directions of principal stress, the stress history, and the strain rate etc. (Atkinson 2000; Chai and Miura 2002; Matasovic and Vucetic 1995; Matsui et al. 1980). Yet, the results obtained by different researchers were often contradictory resulting from the influences of such factors and the test conditions on the dynamic characteristics of soils being beneficial or not, which would improve the laboratory studies and clarify the dynamic behaviors of saturated soft clays.

3.2 Backbone Curve Model

A set of elastic-plastic elements were utilized by Iwan to simulate the dynamic stress-strain backbone curve of soil, based on the theory of kinematic hardening and multiple yielding without considering the yield condition, the flow rule and the boundary conditions, called Iwan model (Iwan 1966; Iwan 1967). The test data can be well fitted by its mathematical expressions.

The strain of each elastic-plastic element is equal during the whole process of loading. However, the stress of each elastic-plastic element is determined by the stiffness of each elastic element and the yield level of each plastic element. If the plastic element does not yield, the load is born by the elastic element and the elastic deformation occurs. But once the plastic element yields, the elastic element cannot afford the load exceeding the yield stress and the stress of the elastic-plastic element is always equal to the yield stress.

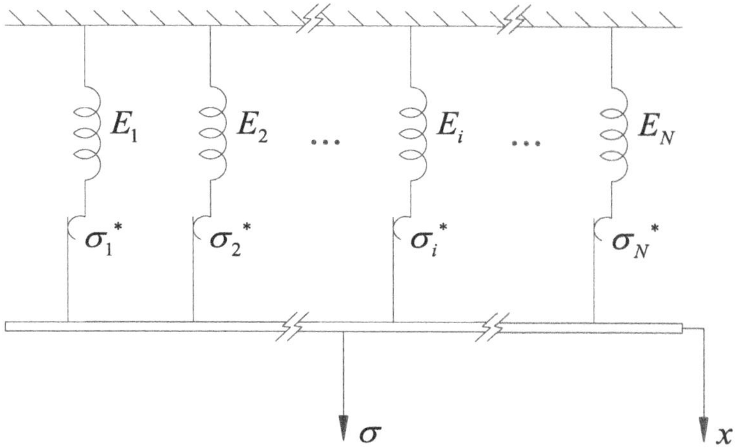

Fig. 3.1 Elastic-plastic elements in the parallel-series Iwan model

The parallel-series Iwan model consists of a collection of N elastic and plastic elements which are called Jenkin's elements, arranged as indicated in Fig. 3.1. The strain ε of each element is equal, and the stress σ is the sum of each individual element. These N elements are sorted and numbered based on the magnitude of their yield strains. For the ith element, E_i is its elastic modulus and σ_i^* is its yield stress. If there are j elements yielding in these N elastic-plastic elements when the model is subjected to load, the relationship between the stress and the strain can be described as follows.

$$\sigma = \sum_{i=1}^{j} \sigma_i^* + \varepsilon \sum_{i=j+1}^{N} E_i \tag{3.1}$$

If E_i of each element equals E, and the total number of elements N becomes very large, Formula (3.1) can be transformed into its equivalent expression.

$$\sigma = \int_0^{E\varepsilon} \sigma^* \phi(\sigma^*) d\sigma^* + E\varepsilon \int_{E\varepsilon}^{\infty} \phi(\sigma^*) d\sigma^* \tag{3.2}$$

where $\phi(\sigma^*)$ is the distribution function of σ^* and $\phi(\sigma^*)d\sigma^*$ is the fraction of the total number of elements having $\sigma^* \leq \sigma \leq \sigma^* + d\sigma^*$. The first term of Formula (3.2) on the right side of the equal sign indicates the stress of elements which have been yielded in the model, and the second term indicates the stress of those which are not yielded when the strain becomes ε.

The derivation of Formula (3.2) is the slope of the stress-strain curve. When $\varepsilon \to 0$, the initial tangent modulus is given by

$$E_0 = E \int_0^\infty \phi(\sigma^*) d\sigma^* \tag{3.3}$$

If the second term of Formula (3.2) vanishes as $\varepsilon \to \infty$, the system will have an ultimate stress σ_{ult} which is given by

$$\sigma_{\text{ult}} = \int_0^\infty \sigma^* \phi(\sigma^*) d\sigma^* \tag{3.4}$$

In this case, it is no longer necessary to obtain the value of the initial tangent modulus. The ultimate stress or the yield force for each particular element is only determined by the distribution function of the individual yield stress.

Almost any function could be used in the foregoing analysis. According to the stress-strain curve in tests, the distribution function can be assumed as

$$\phi(\sigma^*) = \frac{2a^2}{(a + \sigma^*)^3} \tag{3.5}$$

where a is the reciprocal of the ultimate stress. Substituting Formula (3.5) into Formula (3.2), the backbone curve can be obtained as follows.

$$\sigma = \frac{\varepsilon}{\frac{1}{E_0} + \frac{\varepsilon}{\sigma_{\text{ult}}}} \tag{3.6}$$

It can be noted obviously that Formula (3.6) is consistent with Hardin-Drnevich backbone curve model (Hardin and Drnevich 1972).

Amending the Hardin-Drnevich model, Martin-Davidenkov model is obtained (Martin and Seed 1982).

$$\sigma = \left(1 - \left(\frac{(\varepsilon)^{2\text{B}}}{(\varepsilon_0)^{2\text{B}} + (\varepsilon)^{2\text{B}}}\right)^{\text{A}}\right) E_0 \varepsilon \tag{3.7}$$

where ε_0 is the abscissa of the intersection between two tangents passing through the origin and the vertex of the stress-strain curve, respectively.

When A = 1 and B = 0.5, Formula (3.7) can be degraded to Hardin-Drnevich model.

The Hardin-Drnevich and Martin-Darvidenkov hyperbolic models are frequently used to describe the response of soft clay under the dynamic loading. Because of the simple formulation and the clear physical meaning, the parameters can be obtained easily. In order to find a dynamic constitutive model that is suitable for the saturated soft clay in Shanghai, cyclic triaxial tests are conducted to obtain the backbone curve of the soft clay. Then the application and accuracy of the Hardin-Drnevich model and Martin-Darvidenkov model fitting the dynamic stress-strain curve of the soft clay before and after freezing and thawing are analyzed.

Table 3.1 Properties of the silty clay

Soils	Water content (%)	Initial density (g/cm^3)	Void ratio	Liquid limit	Plastic limit	Cohesive strength (kPa)	Angle of internal friction (°)	Permeability coefficient (m/s)
F	36.7	1.78	1.06	38.4	21.2	8.699	28	2.51×10^{-9}
U	35.4	1.81	1.013	36.7	21.3	7.461	30	1.1×10^{-9}

Note F represents freezing-thawing clay; U represents undisturbed clay

3.3 Cyclic Triaxial Tests

3.3.1 Sample Preparation and Soil Properties

The soil samples were taken from the grey silty clay of layer No. 5 typically 5.30–8.40 m in thickness. In order to keep the samples in the undisturbed state, the soil samples were all obtained by thin-walled stainless steel tubes, 100 mm in diameter and 300 mm in length. Each tube was excavated carefully, both ends sealed with wax, transported to the laboratory, and was stored in a humidity room. In this study, some freezing-thawing clay samples are needed. So half of the tubes are put into the freezing box for longer time over 72 h with constant temperature −30°C. Then, the frozen clay is put into the humidors to thaw sufficiently. The properties of the soft clay are summarized in Table 3.1. After freezing and thawing, the porosity of soft clay was increased by 2.25%.

3.3.2 Test Procedures

In this chapter, the sine wave was chosen and one-way stress-controlled cyclic triaxial tests were conducted under consolidation and undrained conditions considering the lower permeability of soft clay.

Ten samples, each with the size of 39.1 mm in diameter and 80 mm in height as the geotechnique test regulation of CNS (China Nation Standard) required, were trimmed from the core of the block sample by a wire saw. Each sample was installed on the base of the apparatus. Sintered bronze end platens were placed between the sample and the apparatus to reduce friction. Drainage of each sample was fulfilled through side draining with the filter paper connected to the bronze platens and the pore pressure was measured through a hole in the center of the rubber sheets covering the lower pedestal.

Each sample was first saturated with the back pressure of 100 kPa and the B value check with the value greater than 0.95 was conducted. Then, each sample was consolidated under equipressure condition for 4 h. After that, the sample was

Table 3.2 Schemes of undrained cyclic triaxial tests

Sl. no	Soils	Confining pressure (kPa)	Axial stress (kPa)	Deviator stress (kPa)	Frequency (Hz)	CSR
S01	F	240	300	60	0.5	0.125
S02	F	240	300	60	0.5	0.25
S03	F	240	300	60	0.5	0.375
S04	F	240	300	60	1.0	0.375
S05	F	240	300	60	2.5	0.375
S06	U	240	300	60	0.5	0.125
S07	U	240	300	60	0.5	0.25
S08	U	240	300	60	0.5	0.375
S09	U	240	300	60	1.0	0.375
S10	U	240	300	60	2.5	0.375

consolidated under K_0 condition for 24 h to perfectly simulate the in-site stress state that was called the initial anisotropy of soil. Because the anisotropic consolidation between the vertical stress and the horizontal stress was produced by the body weight of soil and the horizontal earth pressure, respectively (Li et al. 2011). The parameter K_0 was taken as 0.7 according to the physical and mechanical properties of the grey silty clay and the practical engineering experience. In addition, the confining pressure kept 240 kPa determined by the in-site position where the sample was taken from. The axial cyclic load was applied by changing the amplitude of deviator stress, 30 kPa in the maximum amplitude. Hence, the cyclic stress ratio, CSR $= q_{\mathrm{ampl}}/q_{\mathrm{f}}$, where q_{ampl} is the amplitude of deviator stress and q_{f} is the undrained shear strength obtained from monotonic triaxial tests, which was 80 kPa with the values of freezing-thawing clay and undisturbed clay being 83 kPa and 78 kPa, respectively. Finally, three different frequencies of the vibration loading 0.5 Hz, 1.0 Hz and 2.5 Hz for the cyclic triaxial tests were selected by the measured frequencies induced by the subway loading.

The schemes of the cyclic triaxial tests for the freezing-thawing and the undisturbed soil are summarized in Table 3.2.

3.4 Analysis of Test Results

Clay samples before and after freezing and thawing were prepared, and one-way controlled cyclic triaxial tests were conducted in the laboratory (Cui and Zhang 2015). Dynamic behaviors of the saturated soft clay before and after freezing were studied. The one-way stress-controlled sine wave was selected as the deviator stress whose amplitude was 30 kPa, 0.5 Hz in frequency. As is shown in Fig. 3.2, the typical deviator stress is applied to sample S03, and Fig. 3.2b illustrates its effective stress

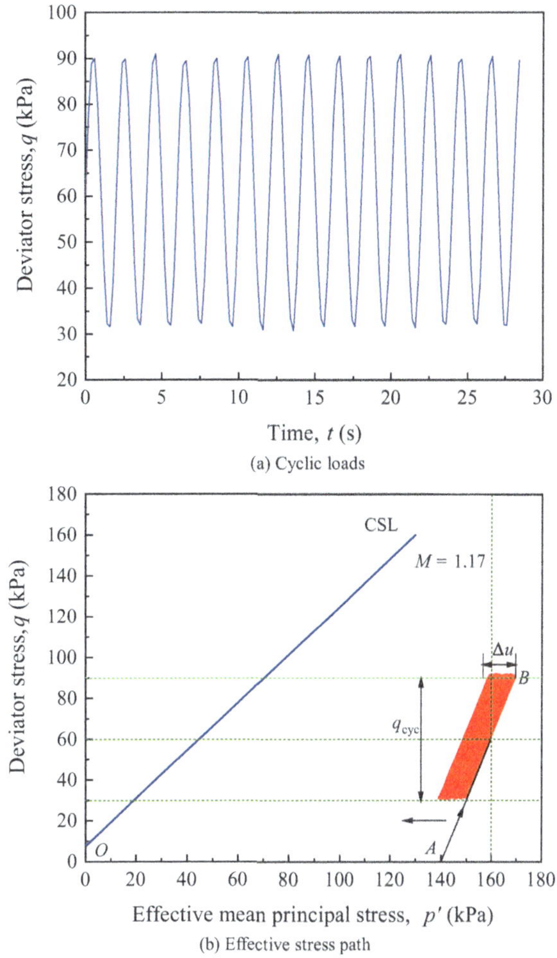

Fig. 3.2 Cyclic stress in cyclic triaxial tests

path. The line from point O to point A represents the stage of isotropic consolidation under drained condition. And the stage of K_0 consolidation is from point A to point B. Cyclic loads was applied after point B. With the increase of the number of cycles, the effective stress path moves close to the CSL (critical state line). Δu is the excess pore water pressure accumulated in the tests.

Based on the test data, the variations of the axial strain and the pore pressure with the number of repeated dynamic loading cycles were studied for the freezing-thawing and the undisturbed silty clay, respectively. Figure 3.3 is the typical result of the axial strain and the excess pore pressure of sample S03. As shown in Fig. 3.3a, the axial strain increases with the number of vibration increasing. The total axial strain induced by the cyclic loading can be divided into the elastic strain and the plastic strain which all increase with the number of vibration increasing. Figure 3.3b shows the excess pore pressure versus time. It can be seen that the excess pore pressure

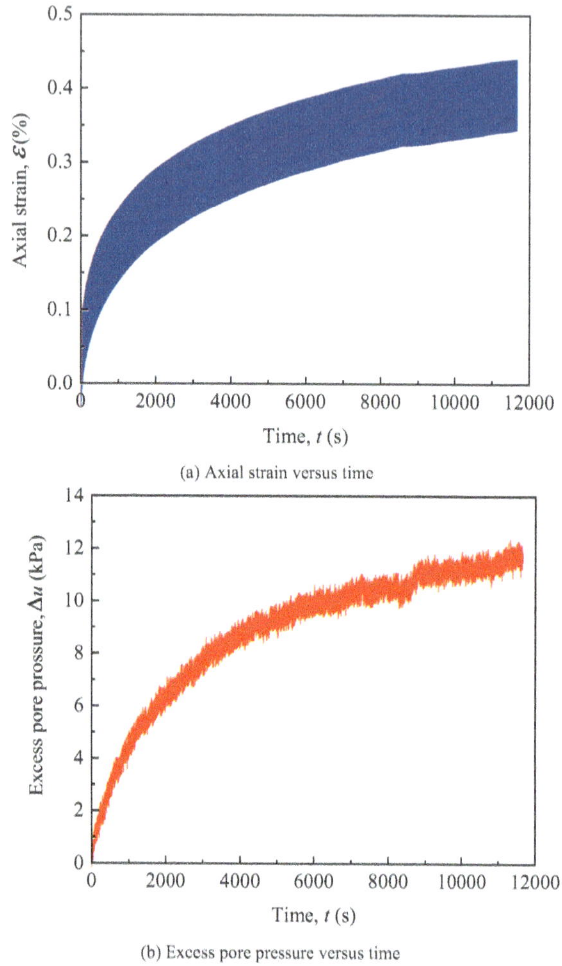

Fig. 3.3 Typical results of cyclic triaxial tests of sample S03

increases sharply during the first 4000 s and then increases slowly in the rest time. Due to the rather low permeability of soft clay, the pore pressure within the samples would not be uniformly distributed. The excess pore water pressure generation during cyclic loading can also be taken into consideration as separate components of the instantaneous and the residual excess pore pressures.

3.4.1 The Stress-Strain Relationship of Soft Clay in Cyclic Tests

Cyclic triaxial tests under the cyclic load with different amplitudes were conducted to the freezing-thawing and the undisturbed samples. The backbone curve was obtained by applying gradual cyclic loading in 20 steps with amplitude of 5 kPa and attaching the peak points of hysteretic curves. As shown in Fig. 3.4, the hysteretic curves of the freezing-thawing and the undisturbed samples with different amplitudes are selected from the first several ones repeatedly and the centers of those are moved to origin in order to reduce the effect of the cumulative residual strain. Then the backbone curves are constructed with the peak points of hysteretic curves. The backbone curves that obtained in tests are fitted by H-D model and M-D model with Formula (3.6) and

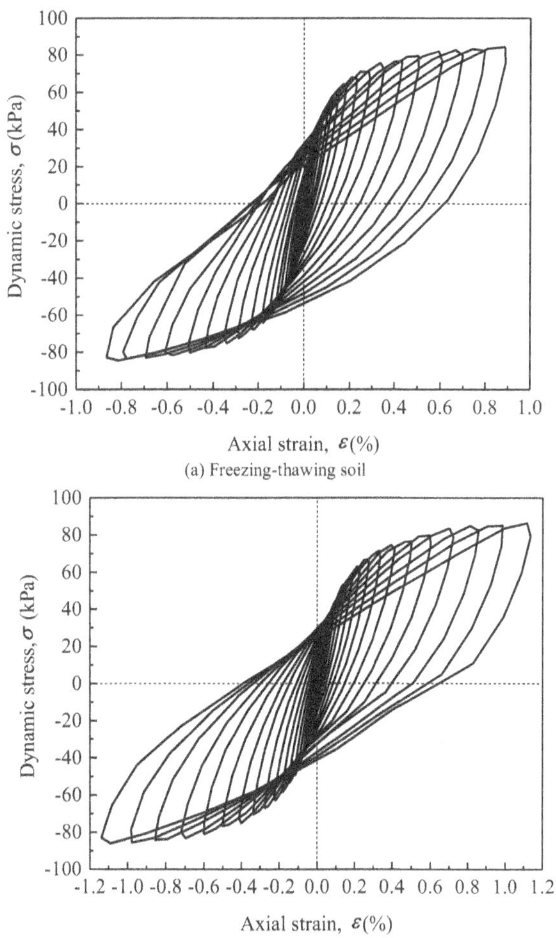

Fig. 3.4 The hysteretic curves of cyclic triaxial tests

Table 3.3 Fitting parameters of two models

Models	Parameters	Freezing-thawing clay	Undisturbed clay
H-D	E_0	1156.904	870.4473
	σ_{ult}	91.78723	94.37236
	R^2	0.999414	0.999026
M-D	A	1.147031	0.600264
	B	0.488894	0.49218
	ε_0	0.064965	0.139589
	E_0	1163.427	1107.052
	R^2	0.999508	0.999457

Formula (3.7), respectively. The fitting parameters of H-D and M-D backbone models are summarized in Table 3.3 for the freezing-thawing samples and the undisturbed samples, respectively.

Figure 3.5a, b illustrates the fitting results of the freezing-thawing and the undisturbed silty clay samples, respectively. It can be seen that both H-D model and M-D model fit the test data well. Comparing with H-D model, M-D model is relatively accurate. Though they both belong to modified hyperbolic models, there are two more parameters of M-D model than that of H-D model. Hence, it is obvious that M-D model agrees with the test data better. The linear correlation coefficients of M-D model are 0.000194 and 0.000431 greater than those of H-D model for the freezing-thawing soil sample and the undisturbed soil sample, respectively.

3.4.2 Comparison of the Axial Strain Before and After Freezing and Thawing

The plastic deformation appears once the cyclic load is applied due to the complex behavior of deformation of the soft clay. And there is no stages for pure elastic deformation throughout the cyclic tests process. The accumulation of the residual plastic strain affects the strength of samples. The values of the elastic strain and the plastic strain have something to do with CSR in cyclic tests (Wang et al. 2013). The increase process of strain can be divided into three obvious stages: the sharply increasing stage, the slowly increasing stage and the smooth stage. Furthermore, the strain has a great relation to soil structures. The more the soil porosity is, the larger deformation the soil produces (Pillai et al. 2010).

Figure 3.6 illustrates the axial strain of samples versus number of vibration with different frequencies. Comparing with the undisturbed samples, the freezing-thawing samples have larger deformation which is about 26.7 and 25.7% than samples with vibration frequency 1.0 Hz and 2.5 Hz, respectively. Though the axial strain of freezing-thawing sample with frequency 0.5 Hz was smaller than that of undisturbed

Fig. 3.5 Fitting results of backbone curves

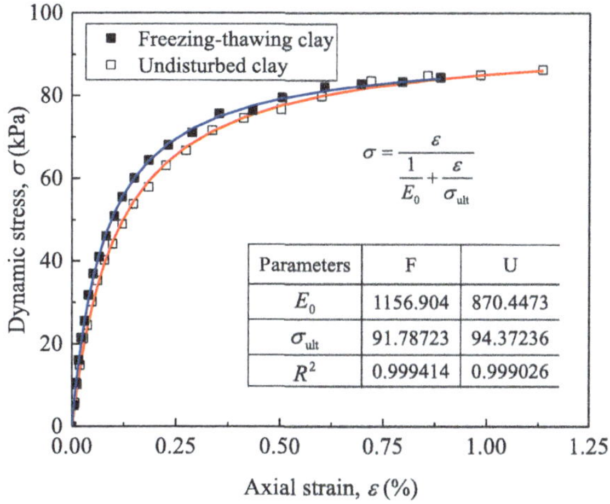

(a) Fitting by H-D

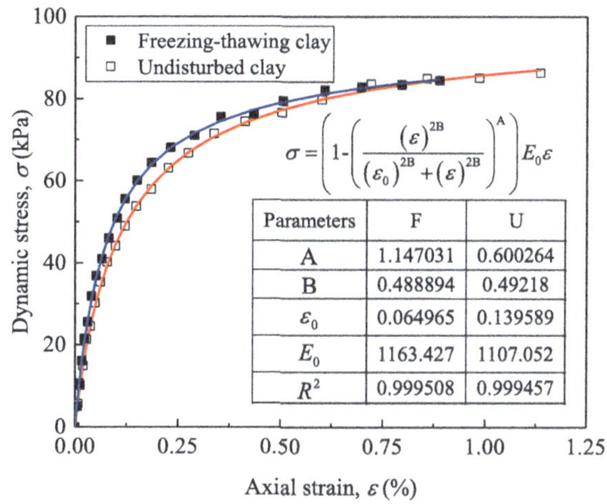

(b) Fitting by M-D

one after 3600 times of vibration, the difference is less than 4%. As discussed before, water in the pore of soil becomes ice with volume expansion after freezing, resulting in the increasing of pore volume within the samples and the undisturbed clay is disturbed. From Fig. 3.6, it can be seen obviously that the frequency has a great effect on the axial strain. Larger axial strain is produced under lower frequency due to the energy of loads transferred to samples. The longer time the loads applied to samples, the more energy that contributes to the deformation is transferred to samples. At the end of tests, the smallest strain is produced with the frequency of 2.5 Hz, while the larger strain is produced with 0.5 Hz.

Figure 3.7 illustrates the axial strain of samples subjected to cyclic load with different amplitudes of deviator stress. Comparing with the undisturbed samples, the

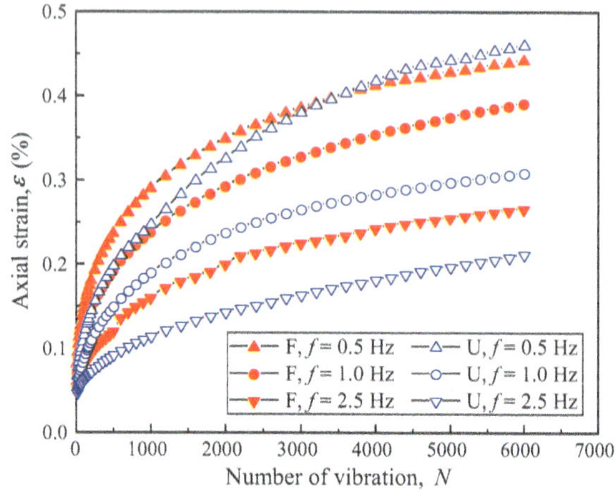

Fig. 3.6 The axial strain of freezing-thawing and undisturbed samples with different frequencies

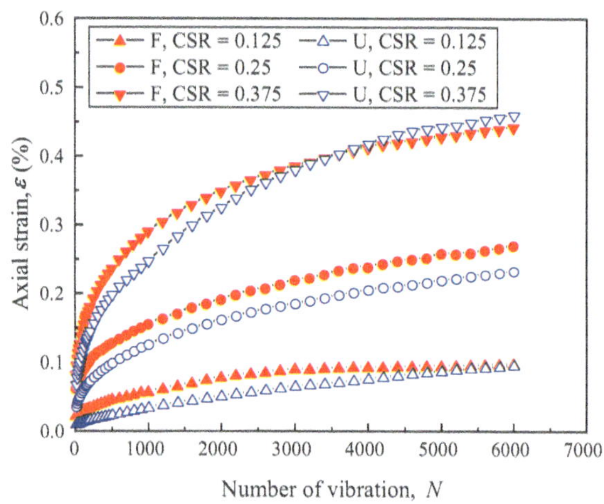

Fig. 3.7 The axial strain of freezing-thawing and undisturbed samples with different amplitudes of deviator stress

samples after freezing and thawing have larger deformation. The CSR values have significant influence on the axial strain. For a low value of CSR, the strain increases rapidly and reaches a comparatively steady state in a short time. However, for a high value of CSR, it takes longer time for the strain increasing and reaching a steady state. It should be considered that the growth rate is reduced with the increasing number of vibration and tending to be stable. The strain of freezing-thawing clay and undisturbed clay increases with the increasing CSR values from 0.125 to 0.375 by 346% and 372%, respectively. With the increase of CSR, the amplitude of the dynamic stress is larger, and the energy transmitted to the samples is more, which results in greater deformation within the samples.

Figures 3.8 and 3.9 illustrate the cumulative plastic strain of samples with different frequencies and CSR in semi-log plots. It can be seen that the plastic strain

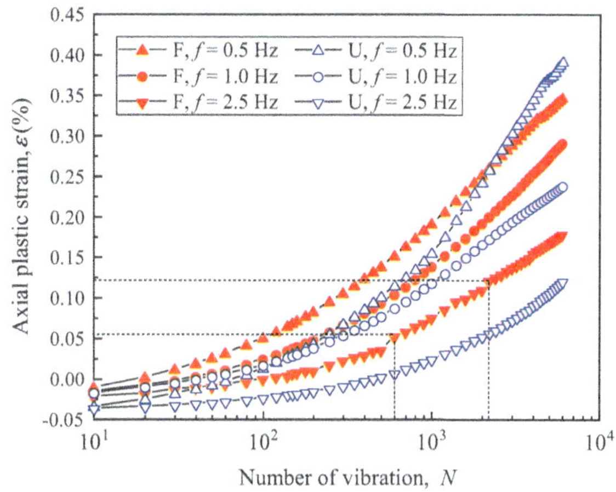

Fig. 3.8 The plastic strain of freezing-thawing and undisturbed samples with different frequencies

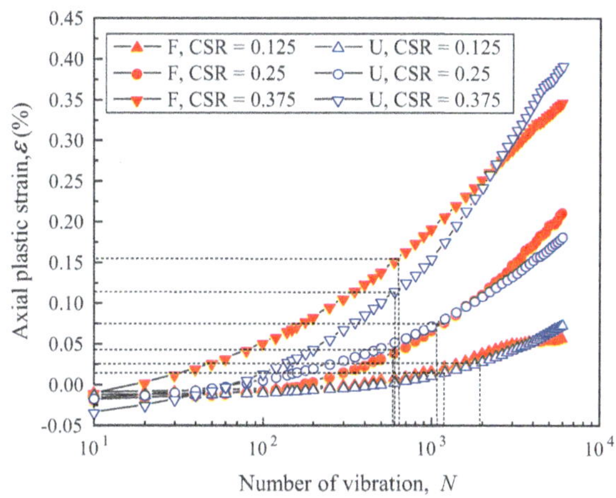

Fig. 3.9 The plastic strain of freezing-thawing and undisturbed samples with different amplitudes of deviator stress

of freezing and thawing samples have larger strain than that of undisturbed samples under the same conditions. This is because cyclic loads aggravate destruction of clay structures after freezing and thawing, which produces a larger deformation relative to undisturbed samples.

Some inflection points can be found from Figs. 3.8 and 3.9. It is apparent that not only the strain values of freezing-thawing samples are larger than that of undisturbed samples, but the inflection points appear earlier. Therefore, inflection points are utilized to recognize the boundary of deformation and destruction. The inflection points that response to the damage of samples structure are not a fixed value, that change with the frequency and CSR. The larger the frequency and the amplitude are, the sooner inflection points appear, and the larger values of the corresponding strain are (Guo et al. 2013). In the initial stage of tests, the plastic strain increases gradually with number of vibration and the strain rate of increasing is not big. But after the increasing of number of vibration, the plastic strain enters a stage of fast growth and tends to destruction with a certain value of the cumulative plastic strain.

3.4.3 The Development of Excess Pore Pressure

The influence induced by cyclic load has been typically attributed to the pore pressure generation. In order to study the behavior of excess pore water pressures of freezing-thawing and undisturbed soft clay more carefully, dynamic cyclic triaxial tests with different frequencies and amplitudes have been conducted to the freezing-thawing and the undisturbed silty clay. With the increase of the number of cycles, excess pore water pressure of clay generated and accumulated, which degrades the structure of clay and decreases the stiffness and strength of clay.

Figure 3.10 shows the development of excess pore water pressures with the increasing of the number of vibration under different frequencies. Comparing with undisturbed samples, the excess pore pressure of freezing-thawing samples increases by 34%, 41% and 53% with 0.5 Hz, 1.0 Hz and 2.5 Hz, respectively. The pore volume of clay is expanded after freezing and thawing. And the pore of clay is all filled with water in saturated samples. According to the effective stress principle, the stress is immediately born by the developing pore pressure when the load is applied. Soil structures are changed directly by the changes of pore pressure under cyclic vibration loads. The lower the frequency, the higher the excess pore water pressure generated. For the frequency of 0.5 Hz, there is longer time for the loads applied to the samples. The longer the samples are subjected to loads, the more energy is transferred to samples, and the higher the excess pore pressure is produced. For the frequency of 2.5 Hz, the time for the loads applied to samples is so short that the excess pore pressure has no time to develop since the energy transferred is little.

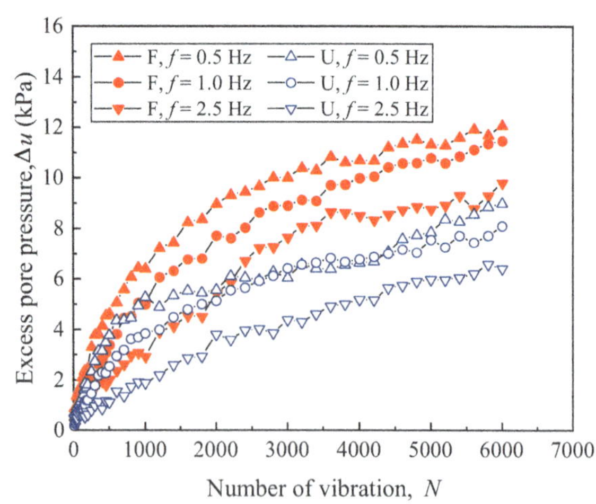

Fig. 3.10 Variations of excess pore pressure of freezing-thawing and undisturbed samples with different frequencies

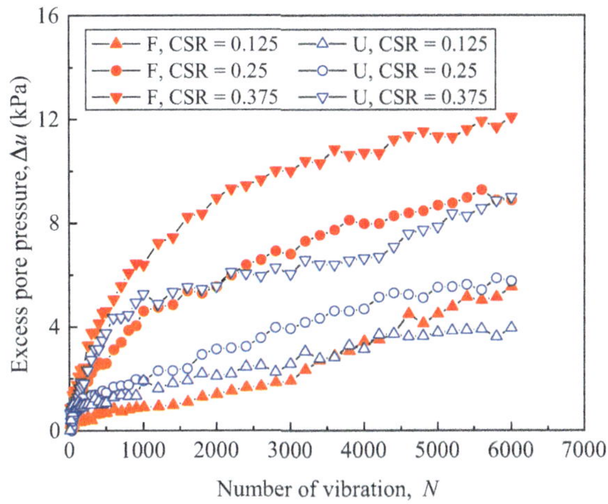

Fig. 3.11 Variations of excess pore pressure responses of undisturbed and freezing-thawing samples with different amplitudes of deviator stress

The development of excess pore water pressures with different CSR values is presented in Fig. 3.11. It can be seen that the freezing-thawing samples have higher pore pressure than that of undisturbed samples. With the same parameters except the CSR, the excess pore pressure increases as the CSR increases. When the CSR is large, the excess pore water pressure increases rapidly at the initial stage of vibration and the curves of the excess pore pressure versus the number of vibration are of characteristic steep slopes. And the increasing rate of excess pore pressure becomes slowly gradually after a certain number of vibration. When the CSR is small, the excess pore pressure begins to develop slowly and gradually. The larger the CSR that represents the amplitude of the dynamic stress is, the more energy transferred to the samples, the longer time is required for the excess pore water pressure to dissipate. It should be pointed out that the excess pore pressure may not distribute uniformly in samples that the excess pore pressure curves are fluctuant due to the complex characteristics of saturated clays.

In order to reflect the relationship between the number vibration and the pore pressure, various empirical models had been established for predicting pore pressure of soil in the previous studies. But there is no one that can be comprehensive for the factors and tests conditions are complex. At present, the most used empirical nonlinear model of excess pore pressure versus log number of vibration can be written as follows:

$$\Delta u = a(\lg N)^2 + b \lg N + c \tag{3.8}$$

where a, b and c are the fitting parameters. From Formula (3.8), there exists the quadratic function relationship between excess pore pressure and the log number of vibration. The fitting results of freezing-thawing and undisturbed samples with different frequencies and CSR values by Formula (3.8) are shown in Fig. 3.12. The fitting parameters are summarized in Table 3.4. The R-Squares are mostly over 0.97, which illustrates that the excess pore pressure model fitted by Formula (3.8) is suitable for the saturated soft clay.

3.5 Conclusions

In this chapter, a series of cyclic triaxial tests have been carried out to investigate the dynamic behavior of soft clays before and after freezing-thawing. The stress-strain backbone loop, axial strain and excess pore pressure of the grey muck clay are evaluated and compared under different frequencies and cyclic stress ratios. The main conclusions are obtained as follows.

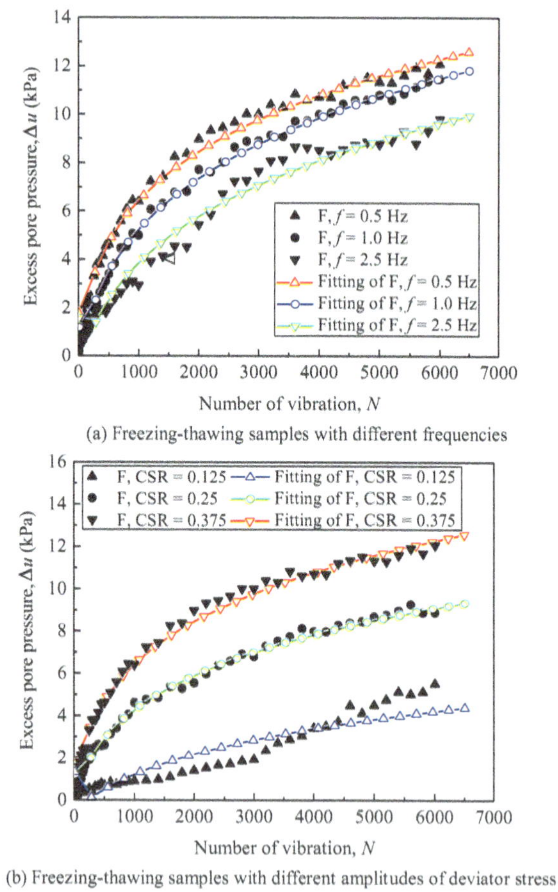

Fig. 3.12 Comparison of curves of experimental data and fitting data

Fig. 3.12 (continued)

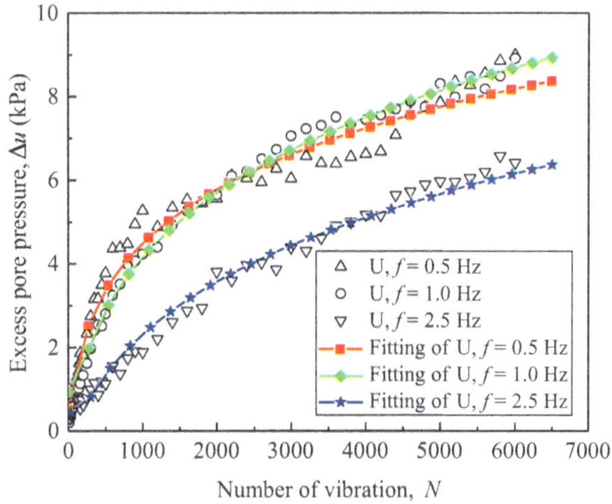

(c) Undisturbed samples with different frequencies

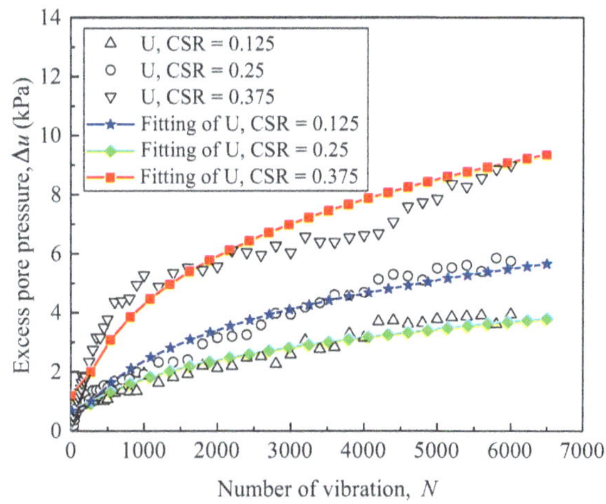

(d) Undisturbed samples with different amplitudes of deviator stress

(1) H-D model and M-D model both belong to the modified hyperbolic Iwan model. M-D model is relatively accurate and agrees better with the test data for freezing-thawing and undisturbed samples comparing with H-D model, even though they both meet the requirements and the results are very close.

(2) Pore water becomes ice with volume expansion during freezing, resulting in the increase of porosity of clay samples. It is equivalent to produce a certain disturbance to the undisturbed clay. Cyclic loads aggravate the destruction of clay structures after freezing and thawing. So the freezing-thawing clay can produce up to 26.7% larger strain and 53% higher excess pore pressure than that of undisturbed clay under the same tests conditions, respectively.

Table 3.4 Fitting parameters by Formula (3.8)

Soils	*a*	*b*	*c*	R^2
F, f = 0.5 Hz, CSR = 0.125	1.57157	−4.47054	3.59371	0.99596
F, f = 0.5 Hz, CSR = 0.25	1.40419	−5.66683	5.63280	0.86734
F, f = 0.5 Hz, CSR = 0.375	1.66419	−3.77024	2.82459	0.99576
F, f = 1.0 Hz, CSR = 0.375	2.06191	−5.96334	4.62044	0.99727
F, f = 2.5 Hz, CSR = 0.375	2.19094	−7.46217	6.59325	0.98168
U, f = 0.5 Hz, CSR = 0.125	1.57157	−4.47054	3.59371	0.99596
U, f = 0.5 Hz, CSR = 0.25	1.13031	−3.67374	3.26187	0.97062
U, f = 0.5 Hz, CSR = 0.375	0.94347	−1.64093	0.94530	0.97877
U, f = 1.0 Hz, CSR = 0.375	1.43541	−3.90192	2.97275	0.99636
U, f = 2.5 Hz, CSR = 0.375	1.58093	−5.79807	5.51303	0.98138

(3) The rates of axial strain and excess pore pressure decrease gradually with the number of vibration increasing. The frequency and the CSR have significant influence on the dynamic behaviors of the saturated soft clay. The lower the frequency and the larger the amplitude of deviator stress are, the more energy transferred to soil is; the larger the axial strain and the higher excess pore pressure are developed, the earlier the inflection points of deformation appear.

References

Atkinson, J. H. (2000). Non-linear soil stiffness in routine design. *Geotechnique, 50*(5), 487–508.

Chai, J., & Miura, N. (2002). Traffic-load-induced permanent deformation of road on soft subsoil. *Journal of Geotechnical and Geoenvironmental Engineering, 128*(11), 907–916.

Chamberlain, E. J., & Blouin, S. E. (1978). Densification by freezing and thawing of fine material dredged from waterways. In *Permafrost 3rd International Conference*. Edmonton, Canada.

Cui, Z.D., & Zhang, Z.L. (2015). Comparison of dynamic characteristics of the silty clay before and after freezing and thawing under the subway vibration loading. *Cold Regions Science and Technology, 119*, 29–36.

Cui, Z. D., He, P. P., & Yang, W. H. (2014). Mechanical properties of a silty clay subjected to freezing–thawing. *Cold Regions Science and Technology, 98*, 26–34.

Eigenbrod, K. D. (1996). Effects of cyclic freezing and thawing on volume changes and permeabilities of soft fine-gained soils. *Canadian Geotechnical Journal, 33*(4), 529–537.

Guo, L., Wang, J., Cai, Y., et al. (2013). Undrained deformation behavior of saturated soft clay under long-term cyclic loading. *Soil Dynamics and Earthquake Engineering, 50*, 28–37.

Hardin, B. O., & Drnevich, V. P. (1972). Shear modulus and damping in soils. *Journal of the Soil Mechanics and Foundations Division, 98*(7), 667–692.

Iwan, W. D. (1966). A distributed-element model for hysteresis and its steady-state dynamic response. *Journal of Applied Mechanics, 33*(4), 893–900.

Iwan, W. D. (1967). On a class of models for the yielding behavior of continuous and composite systems. *Journal of Applied Mechanics, 34*(3), 612–617.

Kokusho, T. (1980). Cyclic triaxial test of dynamic soil properties for wide strain range. *Soils and Foundations, 20*(2), 45–60.

Konrad, J. M. (1989). Physical processes during freeze-thaw cycles in clayey silts. *Cold Regions Science and Technology, 16*(3), 291–303.
Larew, H. G., Leonards, G. A. (1962). A repeated load strength criterion.In *The 41st Annual Meeting of the Highway Research Board*. Washington, D.C.
Lefebvre, G., & Pfendler, P. (1996). Strain rate and preshear effects in cyclic resistance of soft clay. *Journal of Geotechnical Engineering, 122*(1), 21–26.
Li, L., Dan, H., & Wang, L. (2011). Undrained behavior of natural marine clay under cyclic loading. *Ocean Engineering, 38*(16), 1792–1805.
Ling, X., Li, Q., Wang, L., et al. (2013). Stiffness and damping radio evolution of frozen clays under long-term low-level repeated cyclic loading: Experimental evidence and evolution model. *Cold Regions Science and Technology, 86*, 45–54.
Martin, P. P., & Seed, H. B. (1982). One-dimensional dynamic ground response analyses. *Journal of the Geotechnical Engineering Division, 108*(7), 935–952.
Matasovic, N., & Vucetic, M. (1995). Generalized cyclic-degradation-pore-pressure generation model for clays. *Journal of Geotechnical Engineering, 121*(1), 33–42.
Matsui, T., Bahr, M., & Abe, N. (1992). Estimation of shear characteristics degradation and stress-strain relationship of saturated clays after cyclic loading. *Soils and Foundations, 32*(1), 161–172.
Matsui, T., Ito, T., & Ohara, H. (1980). Cyclic stress-strain history and shear characteristics of clay. *Journal of the Geotechnical Engineering Division, 106*(10), 1101–1120.
Pillai, R. J., Robinson, R. G., & Boominathan, A. (2010). Effect of microfabric on undrained static and cyclic behavior of kaolin clay. *Journal of Geotechnical and Geoenvironmental Engineering, 137*(4), 421–429.
Ren, X., Tang, Y., Xu, Y., et al. (2011). Study on dynamic response of saturated soft clay under the subway vibration loading I: Instantaneous dynamic response. *Environmental Earth Sciences, 64*(7), 1875–1883.
Sangrey, D. A., Castro, G., Poulos, S. J., et al. (1978). Cyclic loading of sands, silts and clays. In *The ASCE Geotechnical Engineering Division Specialty Conference*. Pasadena, California.
Seed, H. B., & Chan, C. K. (1966). Clay strength under earthquake loading conditions. *Journal of Soil Mechanics & Foundations Div, 92*(SM2), 53–78.
Simonsen, E., & Isacsson, U. (2001). Soil behavior during freezing and thawing using variable and constant confining pressure triaxial tests. *Canadian Geotechnical Journal, 38*(4), 863–875.
Tang, Y. Q., Cui, Z. D., Zhang, X., et al. (2008). Dynamic response and pore pressure model of the saturated soft clay around the tunnel under vibration loading of Shanghai subway. *Engineering Geology, 98*(3), 126–132.
Wang, J., Guo, L., Cai, Y., et al. (2013). Strain and pore pressure development on soft marine clay in triaxial tests with a large number of cycles. *Ocean Engineering, 74*, 125–132.
Yasuhara, K., Murakami, S., Song, B. W., et al. (2003). Postcyclic degradation of strength and stiffness for low plasticity silt. *Journal of Geotechnical and Geoenvironmental Engineering, 129*(8), 756–769.
Yasuhara, K., Yamanoucm, T., & Hirao, K. (1982). Cyclic strength and deformation of normally consolidated clay. *Soils and Foundations, 22*(3), 77–91.

Chapter 4
Freezing-Thawing on Dynamic Characteristics of Silty Clay

4.1 Introduction

The subway, which boomed in more and more cities, has been considered as a significant way to solve the urban traffic problems. Up to now, more than 14 subway lines are in operation and some new lines are being built in Shanghai. With the rapid development of the subway rail transit, the settlement of soft clay foundation caused by the cyclic loading has been drawn wide attention and has become one of the research interests of soil dynamic mechanics. The subway vibration loading, whose number of vibration often reaches up to hundreds of thousands of times, may result in the decrease or even failure of the stiffness, the strength and the bearing capacity of soil foundations. Therefore, a comprehensive research on the dynamic characteristics of soil under cyclic loading is necessary.

Dynamic characteristics of soil include the dynamic modulus, the modulus degradation, the damping ratio, the dynamic strains and the dynamic excess pore water pressure, etc. Among them, the dynamic modulus and the damping ratio are the most fundamental parameters in the analysis of soil dynamic engineering problems. The shear wave velocity tests were usually conducted to estimate the dynamic modulus and the damping ratio of soil (Hanumantharao and Ramana 2008). However, in most geotechnical investigation programs, the in situ dynamic tests cannot be carried out due to the cost considerations and the lack of specialized personnel. Most of researches are conducted through laboratory tests. Cyclic simple shear tests, resonant column tests, torsional cyclic shear tests and cyclic triaxial tests, which represent different shear strain levels, are regarded as four reliable ways to study the dynamic characteristics of soils. Strain dependent modulus degradation and damping ratio curves are generated through different tests for different kinds of soils.

The constitutive model of soil is often studied by cyclic triaxial tests with different amplitudes of cyclic loading. The stress-strain relationship is presented as a series of hysteresis loops. The backbone curve can be obtained by applying gradual cyclic loading and attaching the peak points of hysteretic curves. Based on the constitutive

Z.-D. Cui et al., *Dynamics of Freezing-Thawing Soil around Subway Shield Tunnels*, https://doi.org/10.1007/978-981-15-4342-5_4

model of soil, the calculation formulas for dynamic shear modulus and damping ratio were established by Hardin and Drnevich (1972a, b). Subsequently, Martin and Seed (1982) improved the constitutive model with two more parameters in order to better describe the dynamic stress-strain relationships of soil. Now it is widely accepted that the dynamic modulus and the damping ratio of soils are functions of dynamic strain under cyclic loading. Idriss et al. (1978) introduced the degradation index, $\delta_N = G_N/G_1$, where G_N is the dynamic shear modulus of each hysteresis loop and G_1 is the dynamic shear modulus of the first loop; the degradation parameter, $t = -\log \delta_N / \log N$, which was also proposed to indicate the relationship between δ_N and N. The damping ratio, the ratio of the damping coefficient to the critical damping coefficient, was proposed by Hardin and Drnevich (1972a, b) as an essential parameter for evaluating the energy absorption capability of soft soil.

However, the development of dynamic modulus varies with the properties of soils. For granular cohesionless soil, the degradation of the dynamic modulus were generally characterized at the strain about 10–4% and the damping ratio was only slightly dependent on the density but not significantly affected by the grain size of the particles (Seed et al. 1986). The development of dynamic elastic modulus with small strain of Nevada sand was studied by Vucetic and Mortezaie (2015) through strain-controlled simple shear constant-volume equivalent-undrained tests with NGI-DSS device. There were inflected points from increasing to decreasing in the curves of dynamic modulus versus dynamic strain, where the dynamic strains were 0.10–0.15%. Okur and Ansal (2007) found that there existed not rising stages but falling stages in the dynamic modulus curves of fine-grained soil samples under undrained conditions. In a single descent curve with the strain lager than 10–3%, two inflected points are regarded as the elastic threshold and the viscoplastic threshold of damage under cyclic loading, respectively.

Dynamic multi-purpose triaxial and torsional shear apparatus were conducted by Luan et al. (2010) to perform undrained cyclic coupling shear tests and cyclic torsional shear tests with hollow cylindrical samples from the saturated soft clay in the Yangtze Estuary. The relationships between the degradation index and the number of vibration were different in two kinds of tests. The model of dynamic modulus degradation can be proposed by the modified Iwan model with regression analysis. The developing models are different with different kinds of soil. Even to the same soil, the result will vary with different test apparatuses and controlled factors. The influence of dynamic amplitude, confining pressure, consolidation ratio, frequency and number of vibration on the dynamic modulus and the damping ratio of soft clay was examined by Li and Wang (2013) through a series of dynamic elasticity modulus tests. The results showed that the dynamic elastic modulus decreased as the dynamic strain developed. The maximum of dynamic modulus of soft clay was proportional to the consolidation ratio, the confining pressure and the frequency. Both the dynamic modulus and the damping ratio increased with the frequency increasing. The modulus degradation with number of vibration versus OCR value was modeled by (Vucetic and Dobry 1988) on consolidated clay and the developing rate of modulus degradation decreased with the OCR increasing. Mortezaie and Vucetic (2013) found that the modulus degradation parameter improved substantially with the frequency and decreased with the vertical consolidation stress.

Besides, freezing-thawing had a great effect on soil dynamic properties. For silty clay, the freezing-thawing process affected the hydraulic conductivity seriously. The soil become looser and the void ratio increased after freezing-thawing, which lead to larger strain under loadings (Cui et al. 2014; Kang and Lee 2015; Zhou and Tang 2015). Freezing-thawing cycles reduced the elastic modulus and enlarged the damping ratio of soils (Wang et al. 2015). The dynamic modulus increased with the freezing temperature descending (Christ et al. 2009). The non-linear behavior of Taipei silty clay was investigated by Lee and Sheu (2007) through a series of undrained cyclic strain-controlled tests and the equations for describing the modulus degradation and the damping ratio were proposed. After freezing, the strength and stiffness of soil would enhance. With the number of vibration increasing, the damping ratio decreased, while the dynamic modulus improved (Ling et al. 2013). Moreover, the elastic modulus changed significantly when the frozen clay melt (Konrad 1989). Zhang and Hulsey (2015) studied the effects of freezing-thawing on soil dynamic properties by using triaxial strain-controlled cyclic tests. The results demonstrated that the shear modulus decreased when the temperature improved from near freezing to above freezing and the damping ratio reached a maximum value when temperature was at or near freezing.

4.2 Definition of Dynamic Modulus and Damping Ratio

In cyclic triaxial tests, a series of continuous hysteresis loops, which gradually move to the right with the increase of number of cycles and dynamic strain, are performed in the dynamic stress-strain curves. Theses closed hysteresis loops of dynamic stress-strain curves are the basis to determine the dynamic modulus and the damping ratio of soil (Kokusho 1980).

4.2.1 Definition of Dynamic Modulus

The equivalent modulus called secant modulus is always utilized to investigate the evolution of dynamic modulus of soil, including the dynamic elastic modulus E and the dynamic shear modulus G. There is a mutual transformation relationship between E and G as shown in Eq. (4.1).

$$G = \frac{E}{2(1 + \mu)} \tag{4.1}$$

where μ is the Poisson's ratio of soil.

As presented in Fig. 4.1, the dynamic secant modulus is determined by the slope of the straight line connecting the two extremes of the hysteresis loop under the ideal condition.

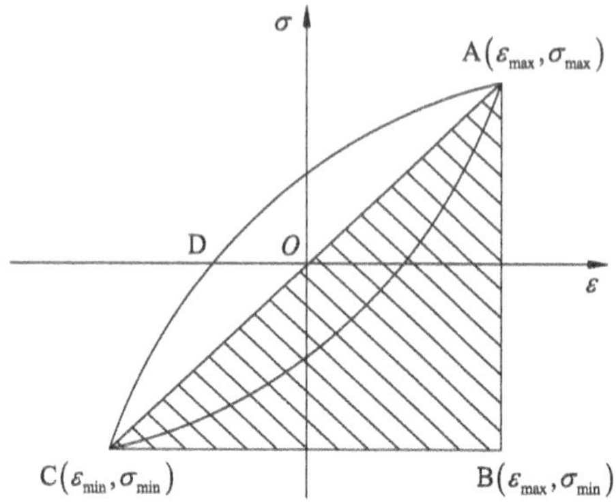

Fig. 4.1 The hysteresis loop under ideal condition

$$E = \frac{\sigma_{max} - \sigma_{min}}{\varepsilon_{max} - \varepsilon_{min}} \tag{4.2}$$

where σ_{max} and σ_{min} are the maximum and minimum stress in a single hysteretic curve, respectively; ε_{max} and ε_{min} are the strains responding to the maximum and minimum stress, respectively.

4.2.2 Definition of Damping Ratio

The damping ratio can represent the capability of soil to absorb energy under cyclic loading. The magnitude of damping can be represented by the damping coefficient, which is the ratio of the damping force to the vibration velocity. In addition, it can also be represented by damping ratio λ, which is the ratio of the actual damping coefficient c to the critical damping coefficient c_{cr}. In cyclic triaxial tests, it can be calculated through the areas enclosed by hysteresis curves in the loading and unloading stages. Accordingly, the damping ratio can be characterized as

$$\lambda = \frac{1}{\pi}\frac{\Delta A}{A} \tag{4.3}$$

where ΔA is the area enclosed by a single hysteretic loop and A is the area of triangular ΔABC as shown in Fig. 4.1.

With deviator stress, the plastic strain of samples will develop and accumulate, resulting in the unclosed hysteretic loops under cyclic loading. Though the plastic strain is little enough to ignore in a single cycle, the accumulated strains are generated. In Fig. 4.1, there is misalignment between point A and C. The damping of samples is undoubtedly associated with the plastic strain due to the properties of silty clay. However, the damping ratio can be determined by the hysteretic curves. Masing (1926) proposed a hypothesis, which was called Masing's double rule, to investigate the shape of elastic-plastic elements. The dynamic stress-strain model is obtained by

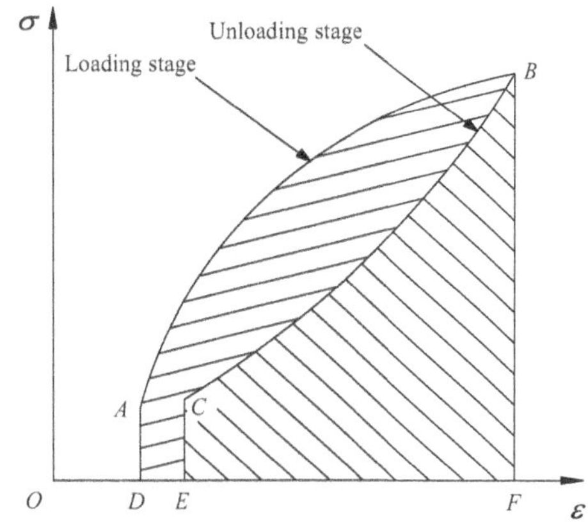

Fig. 4.2 The energy relations between the loading stage and the unloading stage

means of backbone curves, which are utilized to build the relationships of hysteretic curves. Once the backbone curve is determined, the hysteretic curves are obtained in that both of them are in the same curve shapes.

In actual, there are few practical evidences from tests for that hypothesis. The problem of Masing's double rule was that the rule itself had serious limitations in describing the undrained shear behavior of soft clay (Puzrin et al. 1995). The hysteresis loops obtained through Masing's double rule are usually too large. The damping ratio changes all the time due to the complex dynamic behaviors of silty clay, which Masing's double rule cannot simulate. Moreover, there are always deviations between the theory and tests, resulting in the errors of accumulated strains and damping ratios. Therefore, Masing's double rule was banished in this chapter assuming that there was no close relationship between hysteretic curves and backbone curves. The geometric hyperbolic functions written as Eqs. (4.4) and (4.5) were utilized to fit the hysteretic curves. The areas shown in Fig. 4.2 were calculated and the capability of samples to absorb energy was evaluated under different conditions.

For the loading stage, the curve AB can be written as

$$\sigma = \frac{a_i - \varepsilon}{b_i \varepsilon + c_i} \tag{4.4}$$

where a_i, b_i and c_i are the fitting parameters.

For the unloading stage, the curve BC can be written as

$$\sigma = \frac{a_j + \varepsilon}{b_j \varepsilon + c_j} \tag{4.5}$$

where a_j, b_j and c_j are the fitting parameters.

It is proved that the geometric hyperbolic functions fit well with the hysteretic curves. The integral equation of Eq. (4.4), which represents the area enclosed by the loading curve AB and axis of abscissae, can be written as

$$\Delta A_1 = -\frac{\varepsilon}{b_i} + (\frac{a_i}{b_i} + \frac{c_i}{b_i^2}) \ln(\varepsilon + \frac{c_i}{b_i})\Bigg|_A^B \quad (4.6)$$

The area enclosed by the unloading curve BC and axis of abscissae, can be written as

$$\Delta A_2 = \frac{\varepsilon}{b_j} + (\frac{a_j}{b_j} - \frac{c_j}{b_j^2}) \ln(\varepsilon + \frac{c_j}{b_j})\Bigg|_C^B \quad (4.7)$$

The areas of hysteretic loops can be written as

$$\Delta A = \Delta A_1 - \Delta A_2 \quad (4.8)$$

Substituting Eqs. (4.8) into (4.3), the damping ratio of soils under the cyclic loading can be obtained.

The misalignment between point A and point C in Fig. 4.2 represents the plastic strain between the loading stage and the unloading stage of the hysteretic loop. The inclination of line AB represents the change of secant modulus and the level of strain softening within the samples. The bigger the angle between line AB and axis of abscissae is, the bigger the secant modulus is and the lower the level of strain softening is. The bigger the damping is, the higher the level of viscosity and the better the anti-vibration performance is. The calculation method of damping ratio with hyperbolic fitting is simple and the results can reach the requirements, especially the sinusoidal cyclic loading. In this chapter, the cyclic triaxial tests were conducted to investigate the dynamic stress-strain relationship and evaluate the capability of silty clay samples to absorb energy under vibration loading.

4.3 Cyclic Triaxial Tests

4.3.1 The Schedule of Triaxial Tests

The samples were taken from the grey silty clay of layer No. 5 typically 5.30–8.40 m in thickness in Shanghai. In order to keep these samples in undisturbed state, they were obtained by thin-walled stainless steel tubes, 100 mm in diameter and 300 mm in length. Each tube was excavated carefully, both ends sealed with wax, transported to the laboratory and stored in a humidity room. In this study, some freezing-thawing clay samples were demanded. Some tubes were put into the freezing box for time over 72 h with three different constant freezing temperatures, −30 °C, −20 °C and −10 °C, respectively. Then, these tubes were put into the humidors to thaw sufficiently. The properties of the undisturbed silty clay are summarized in Table 3.1. After freezing-thawing, the porosity of silty clay increased by 2.25%.

4.3.2 The Process of Triaxial Tests

The triaxial tests were conducted with the GDS (Global Digital Systems) apparatus displayed in Fig. 4.3a. In this chapter, the sine wave was chosen and one-way stress-controlled cyclic triaxial tests were conducted under consolidation and undrained conditions considering the lower permeability of soft clay (Zhang and Cui 2018).

Twelve samples, each with the size of 39.1 mm in diameter and 80 mm in height, were trimmed from the core of the block sample by a wire saw. Each sample was installed on the base of the apparatus as Fig. 4.3b displayed. Sintered bronze end platens were placed between the sample and the apparatus to reduce friction. Drainage of each sample was fulfilled through side draining with the filter paper connected to the bronze platens and the pore pressure was measured through a hole in the center of the rubber sheets covering the lower pedestal.

Each sample was originally saturated with the back pressure of 100 kPa and B value check with the value greater than 0.95 was conducted. After that, each sample was consolidated under equipressure condition for 4 h. After that, the sample was consolidated under K_0 condition for 24 h to perfectly simulate the in-site stress state that was called the initial anisotropy of soil. The parameter K_0 was taken as 0.7 according to the physical and mechanical properties of the grey silty clay and the practical engineering experience. In addition, the confining pressure kept 240 kPa determined by the in-site position. The axial cyclic load was applied by changing the amplitudes of deviator stress assumed as 10 kPa, 20 kPa and 30 kPa, respectively. Hence, the cyclic stress ratio, CSR $= q_{\text{ampl}}/q_{\text{f}}$, where q_{ampl} is the amplitude of deviator stress and q_{f} is the undrained shear strength obtained from monotonic triaxial tests, being 80 kPa. The CSRs were determined as 0.125, 0.25 and 0.375, respectively. Finally, three different frequencies of the vibration loading 0.5 Hz, 1.0 Hz and 2.5 Hz for the cyclic triaxial tests were selected by the measured frequencies induced by the subway loading (Tang et al. 2008).

The schemes of the cyclic triaxial tests for the freezing-thawing and the undisturbed silty clay are summarized in Table 4.1.

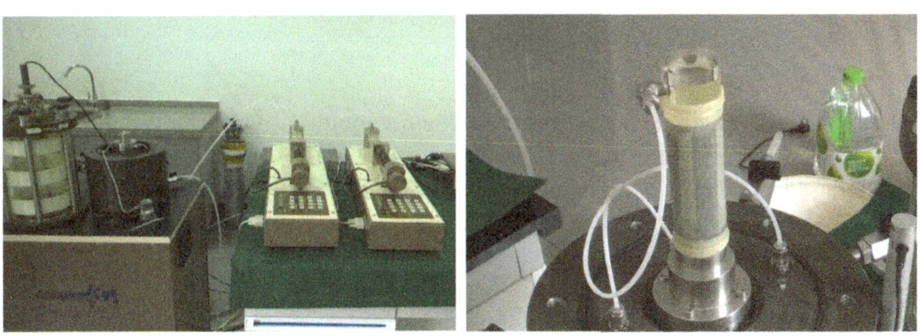

(a) The apparatus (b) The sample

Fig. 4.3 The GDS triaxial tests apparatus and the soil sample installed on the apparatus

Table 4.1 Schemes of undrained cyclic triaxial tests

Sl. no.	Soils	Confining pressure (kPa)	Axial stress (kPa)	Frequency (Hz)	CSR	Temperature (°C)
01	F	240	300	0.5	0.125	−30
02	F	240	300	0.5	0.25	−30
03	F	240	300	0.5	0.375	−30
04	F	240	300	1.0	0.375	−30
05	F	240	300	2.5	0.375	−30
06	F	240	300	0.5	0.375	−20
07	F	240	300	0.5	0.375	−10
08	U	240	300	0.5	0.125	–
09	U	240	300	0.5	0.25	–
10	U	240	300	0.5	0.375	–
11	U	240	300	1.0	0.375	–
12	U	240	300	2.5	0.375	–

4.4 Results Analysis

In this study, clay samples before and after freezing-thawing were prepared, and one-way controlled cyclic triaxial tests were conducted in the laboratory. Due to the hysteretic nature of saturated soft clay of layer No. 5 in Shanghai and the generation of accumulated excess pore water pressure, the stiffness degradation occurs under the cyclic loading. In order to investigate the dynamic properties of silty clay before and after freezing-thawing, especially the dynamic modulus and the damping ratio, 12 cyclic triaxial tests were conducted and the effects of the frequency, the number of vibration, the CSR and the freezing temperature on the dynamic modulus and the damping ratio were discussed.

4.4.1 Dynamic Modulus

Figure 4.4 illustrates the development of dynamic modulus of freezing-thawing and undisturbed samples with different frequencies. It can be seen obviously that the developing laws of secant modulus are complex. For the freezing-thawing samples, the modulus curves decrease within the initial 2000 cycles. Then each curve is fluctuated in a relative stable range. However, for undisturbed samples, the curves decrease within the initial 1000 cycles. The dynamic modulus reduces after freezing-thawing under the same condition, which indicates the significant effect of freezing-thawing on the structures and the strength of silty clay. The dynamic modulus of the freezing-thawing sample with frequency of 0.5 Hz is 11.5% and 14.2% larger than that of 1.0 Hz and 2.5 Hz at the end of the tests, respectively. The lower the frequency is, the

Fig. 4.4 The dynamic modulus of freezing-thawing and undisturbed samples with different frequencies

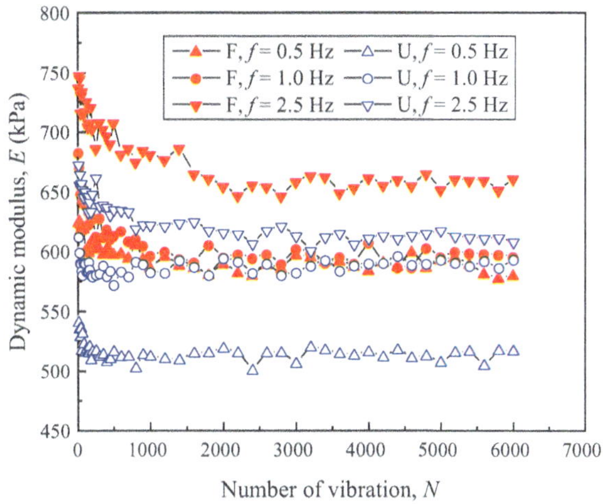

more the energy is transferred to soil. With the strain and the excess pore pressure developing, the bigger degradation appears in samples. In addition, it is obviously that the dynamic modulus of the freezing-thawing samples is larger than that of the undisturbed samples, which is opposite to other research findings mentioned above. It is for the reason that the ice expansion of pore volume results in squeezing of the particle aggregates of silty clay with low permeability. After freezing-thawing, the particle aggregates of the silty clay still remain certain freezing strength, though the structure become loose due to the loss of water.

It can be seen from Fig. 4.5 that the dynamic modulus of freezing-thawing and undisturbed samples increase with the amplitudes of deviator stress increasing. The larger the CSR is, the greater the dynamic modulus reduces under the same conditions. It is related to the energy absorbed by samples under the cyclic loading. The

Fig. 4.5 The dynamic modulus of freezing-thawing and undisturbed samples with different CSRs

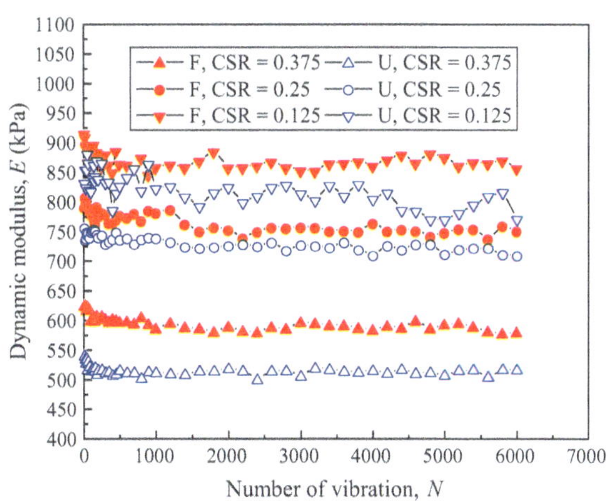

larger plastic strain and higher excess pore water pressure of samples develop under larger deviator stress. However, the increasing energy decreases the stability of samples. Kokusho (1980) reported that the stiffness of soils did not change significantly with number of vibration when the strain amplitude was small, while the modulus would decrease significantly if the strain was large for the development of the pore water pressure. However, the results of silty clay samples are opposite in this study. The modulus decreases significantly when the strain is small for the reason that the samples are not compacted at the initial stage. With the number of vibration increasing, the increment of plastic deformation and excess pore water pressure become smaller and smaller. The samples begin to harden and easily break with the energy continuously absorbed, though the modulus changes slightly at the end of tests.

Figure 4.6 presents the development of dynamic modulus of freezing-thawing and undisturbed samples with different freezing temperatures. The dynamic modulus of freezing-thawing samples with freezing temperatures of −10 °C and −20 °C is larger than that of −30 °C. The modulus of the freezing-thawing sample with −30 °C is 12.2% larger than that of the undisturbed sample under the same conditions. In addition, the axial strain of sample with freezing temperature of −10 °C is larger than that of −20 °C. When the temperature drops to −10 °C in the freezing process, the water within the pore fully becomes ice with volume expanding. The disturbance from ice extrusion to clay structures is the most powerful after thawing. However, when the temperature drops to −30 °C, the clay structures are strengthened conversely with the developed ice compacting soil particles. Thus, −10 °C can be regard as an inflected point of freezing temperature that determines the dynamic characteristics of freezing-thawing clay. And it is consistent with the results of uniaxial compression tests conducted by Christ et al. (2009) with silty soil to analyze the effects of freezing-thawing cycles on the dynamic modulus.

For the samples under the gradual cyclic loading, the backbone curve can be obtained by attaching the peak points of hysteretic curves. But for samples with

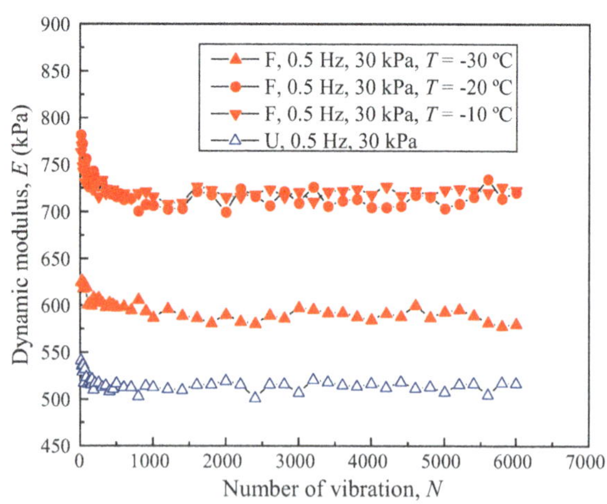

Fig. 4.6 The dynamic modulus of freezing-thawing and undisturbed samples with different freezing temperatures

the equal amplitude of dynamic stress, there is no close relationship between the backbone curve and hysteretic curves (Seed and Idriss 1970). In this chapter, the hyperbolic functions are utilized to fit the hysteretic curves and the dynamic modulus can be expressed as a function of dynamic strain with three parameters as follows:

$$E = \frac{A}{1 + B\varepsilon^{C}} \tag{4.9}$$

where A, B and C are the fitting parameters of tests.

Figure 4.7 illustrates the fitting results of modulus reduction versus dynamic strain of silty clay subjected to the cyclic loading. It can be seen obviously that the dynamic modulus is reduced significantly after freezing-thawing. Moreover, the dynamic modulus increases with the frequency decreasing and the CSR increasing, while the axial strain decreases. Of course, the range of strain in cyclic triaxial tests is much larger than that in resonant volume tests, which results in vacancies while the strain approaches to zero. The fitting parameters are summarized in Table 4.2 and the results meet the requirements.

4.4.2 Damping Ratio

The damping ratio of soil represents the capability of samples to absorb energy under cyclic loading. In actual, the energy transformed into soil partly leads the plastic deformation of samples and partly increases the excess pore pressure under the cyclic loading. The more the energy is absorbed by samples, the larger the damping ratio is. Figures 4.8, 4.9 and 4.10 illustrate the variations of the damping ratio with number of vibration under different frequencies, CSRs and freezing temperatures. The damping ratio is fluctuating and decreases with the increase of number of vibration in general. The reason may be that the samples belong to saturated silty clay, which are characterized by high water content, high compressibility and low permeability. The excess pore water pressure increases quickly under the cyclic loading, while its dissipation process takes long time. It should be noted that the energy consumed to increase the excess pore pressure occupies most of the damping consumed energy. Therefore, the fluctuation of excess pore water pressure leads to the fluctuant damping ratio (Lee and Sheu 2007; Liu and Kai 2010).

Figure 4.8 shows that the damping ratio of samples with the frequency of 1.0 Hz experiences the maximum, and the damping ratios are nearly same before and after freezing-thawing. In fact, the frequency of 1.0 Hz is close to the eigen frequency of samples in this study, which may cause resonance effect on samples and lead to lager deformation. Once the resonance effects reach the strength of the particle aggregates, the behaviors of the silty clay will exhibit no regularity, e.g., the damping ratio of freezing-thawing samples with frequency of 2.5 Hz is larger than that before freezing-thawing. In general, the damping ratios of both the freezing-thawing sample and the undisturbed sample under the same frequency are similar, which indicates

Fig. 4.7 The fitting results of dynamic modulus versus axial strain

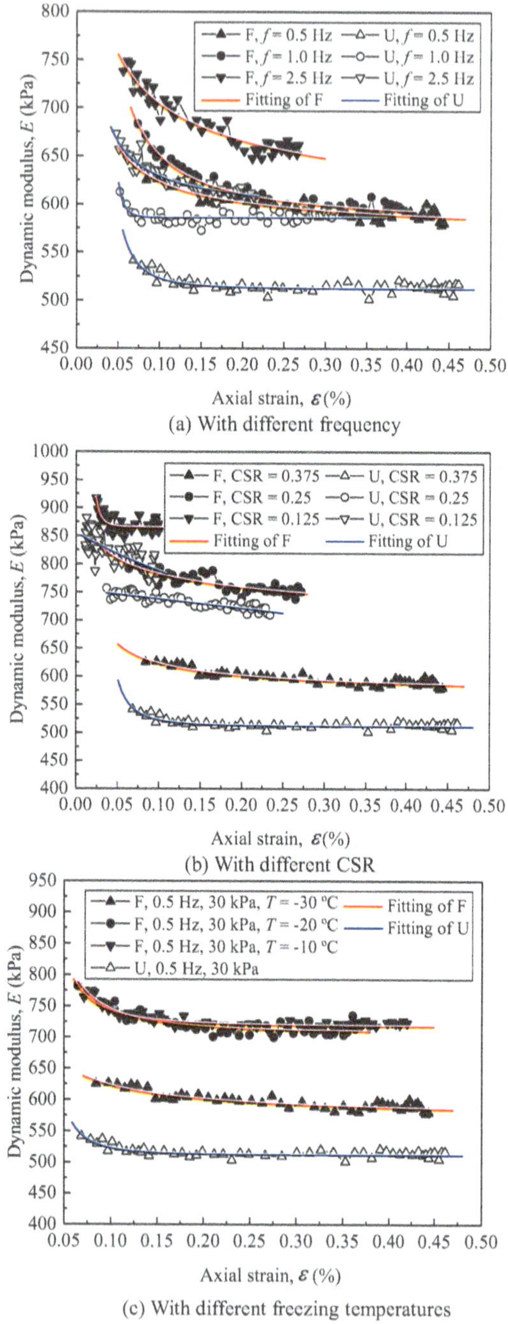

Table 4.2 Fitting parameters by Eq. (4.9)

Soil	*A*	*B*	*C*	R^2
F (T = −30 °C), f = 0.5 Hz, CSR = 0.125	865.7215	−4.7E−09	−4.36278	0.7548
F (T = −30 °C), f = 0.5 Hz, CSR = 0.250	500.4508	−0.29894	−0.08206	0.8696
F (T = −30 °C), f = 0.5 Hz, CSR = 0.375	563.9918	−0.02427	−0.59747	0.8968
F (T = −30 °C), f = 1.0 Hz, CSR = 0.375	577.7052	−0.01015	−1.04687	0.9706
F (T = −30 °C), f = 2.5 Hz, CSR = 0.375	537.1032	−0.12147	−0.29175	0.9615
F (T = −20 °C), f = 0.5 Hz, CSR = 0.375	717.9345	−0.00031	−2.05441	0.8851
F (T = −10 °C), f = 0.5 Hz, CSR = 0.375	705.9863	−0.00133	−1.60414	0.9076
U, f = 0.5 Hz, CSR = 0.125	857.6508	0.67712	0.89363	0.7193
U, f = 0.5 Hz, CSR = 0.250	755.4879	0.25087	1.06975	0.8606
U, f = 0.5 Hz, CSR = 0.375	512.9727	−3.8E−05	−2.73477	0.8004
U, f = 1.0 Hz, CSR = 0.375	586.7493	−7.6E−12	−7.60607	0.7181
U, f = 2.5 Hz, CSR = 0.375	585.1632	−0.01264	−0.75058	0.9493

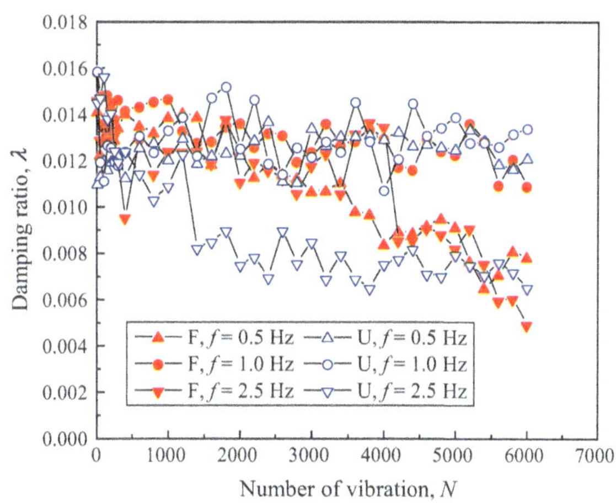

Fig. 4.8 The damping ratio of freezing-thawing and undisturbed samples with different frequencies

that freezing-thawing has slight effect on the damping properties of silty clay. With the increasing of number of vibration, the increment of deformation and excess pore water pressure within samples under the cyclic loading become smaller and smaller. Though the energy absorbed by samples decreases in each cycle, the damage of soil structures is more and more serious and the samples break more and more easily.

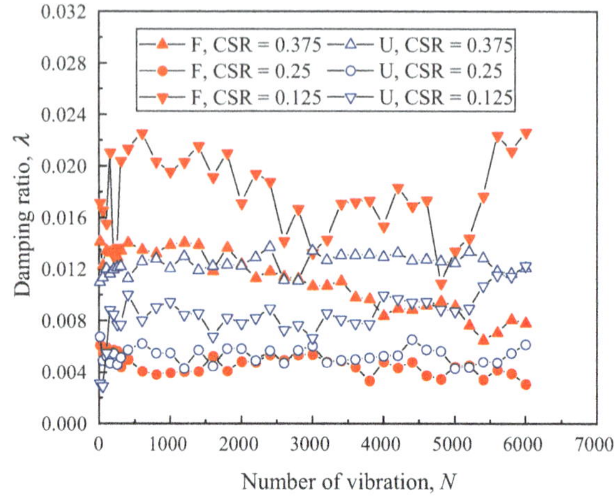

Fig. 4.9 The damping ratio of freezing-thawing and undisturbed samples with different CSRs

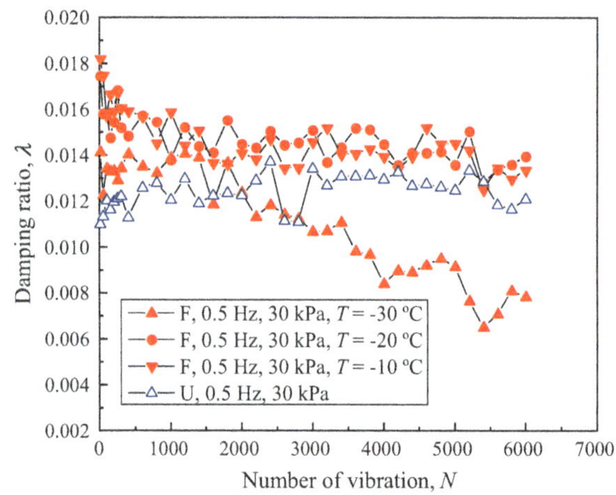

Fig. 4.10 The damping ratio of freezing-thawing and undisturbed samples with different freezing temperatures

Figure 4.9 illustrates the variations of the damping ratio of silty clay with number of vibration under the cyclic loading. For the freezing-thawing samples, the damping ratio experiences the maximum with CSR of 0.125 and the one with CSR of 0.375 takes the second place. For the undisturbed samples, it is opposite. According to the cyclic triaxial tests in this chapter, the amplitudes of deviator stress keep constant in the whole process. In theory, the plastic strain consumes more energy and the damping ratio increases with the CSR increasing, which deviates from the tests results. The reasons may be that there are misalignments within the samples. Meanwhile, the deformation of samples is bidirectional including both the axial direction and the radial direction. Though the axial strain occupies a dominant position, ignoring the radial strain is not rigorous.

It can be seen from Fig. 4.10 that all the damping ratio curves decrease with the number of vibration increasing. For the sample with temperature of −30 °C, the dropping rate achieved about 50%. In addition, the curves are close to each other overall, especially in the first half part. Under the vibration loading, the deformation and pore water pressure develop in samples. However, the increments of deformation and the pore water pressure decrease gradually due to the hardening of soil. The lower the temperature is, the greater the pore volumes expand and more room is obtained for deformation of samples after thawing (Ling et al. 2009). The decline of damping ratio before and after freezing-thawing increases with the freezing temperature dropping. It should be mentioned again that the magnitudes of strain measured in cyclic triaxial tests are larger than those in resonant column apparatus.

4.4.3 Excess Pore Water Pressure and Axial Strain

Figure 4.11 presents the development of the excess pore water versus number of vibration with different frequencies, amplitudes of deviator stress and freezing temperatures, respectively. Comparing with undisturbed samples, the excess pore pressure of freezing-thawing samples increases by 34%, 41% and 53% with 0.5 Hz, 1.0 Hz and 2.5 Hz, respectively. When the CSR increases 2 times from 0.125 to 0.375, the excess pore water pressure of freezing-thawing and undisturbed samples increases about 1.2 and 1.3 times, respectively. It should be mentioned that the excess pore pressure may not distribute uniformly in samples and the excess pore pressure curves are fluctuant due to the complex characteristics of saturated silty clay.

Figure 4.11a illustrates that the excess pore water pressure decreases with the increasing of frequencies. The excess pore water pressures of freezing-thawing and undisturbed samples with 0.5 Hz are about 23% and 40% bigger than those with 2.5 Hz, respectively. The smaller the frequency of dynamic stress is, the more energy is transferred to samples and the bigger the pore water pressure is. Figure 4.11b shows that excess pore water pressure increases with the amplitudes of deviator stress increasing. When the CSR is small, the excess pore pressure begins to develop slowly and gradually. When the CSR is large, the excess pore water pressure increases rapidly at the initial stage of vibration and the increasing rate of excess pore pressure becomes slowly gradually after about 1000 times of vibration. The larger the CSR is, the more energy is consumed and the longer the time is required for the excess pore water pressure to dissipate. It can be seen from Fig. 4.11c that there are no obvious differences among the excess pore pressures of freezing-thawing samples with freezing temperatures of −10 °C, −20 °C and −30 °C, respectively. The maximum is 13 kPa approximately, which is improved about 44% compared with that of the undisturbed sample under the same conditions. Though the gaps of excess pore water pressure among freezing-thawing samples are small, the excess pore water pressure increases with the increasing of freezing temperature for the reason that the pore volumes within samples expand after freezing-thawing.

Fig. 4.11 Excess pore water pressure of undisturbed and freezing-thawing samples

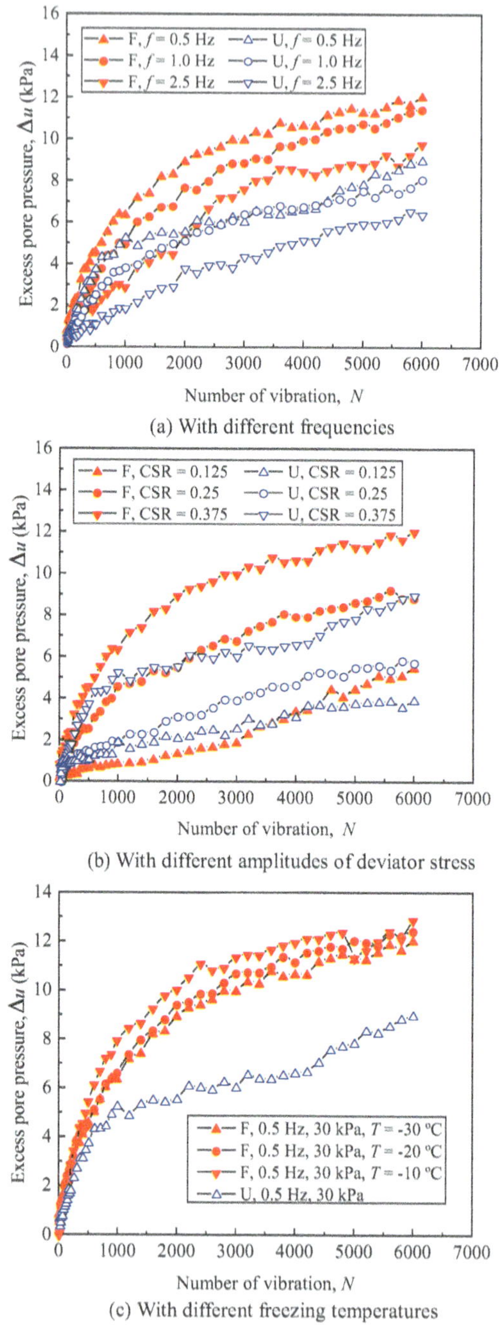

The development and dissipation of excess pore water pressure disturb the structures of clay and result in deformation. The development of axial strains with different frequencies, CSR values and freezing temperatures is presented in Fig. 4.12. The pore volume is expanded after freezing-thawing, which is equivalent to a certain disturbance to the structures of samples, resulting in bigger axial strain after freezing-thawing. The axial strains increase with the number of vibration increasing.

Figure 4.12a illustrates the development of axial strain of silty clay samples with different frequencies. It can be seen that frequencies of cyclic loading have a great effect on the behaviors of axial strain. For the samples under the same conditions, the lower the frequency is, the bigger the axial strain is (Cui and Zhang 2015). The axial strains of freezing-thawing and undisturbed samples with 0.5 Hz are 65.8% and 116.7% larger than those with 2.5 Hz, respectively. Figure 4.12b shows the development of axial strain with number of vibration under different amplitudes of deviator stress. It is obvious that the axial strain increases with the CSR increasing. Except the last part of curves with the CSR of 0.375, the axial strains of freezing-thawing samples are always lager than those of undisturbed samples. Under the same tests conditions, the axial strains of samples with CSRs of 0.25 and 0.125 increase 16% and 5% after freezing-thawing, respectively. Figure 4.12c illustrates the axial strain developing with different freezing temperatures. For the sample with the freezing temperature of −30 °C, the axial strain is the largest though the last part of its curve is lower than that of the undisturbed sample. The reason for that might be that the properties of samples differ from each other or there might be uncontrollable factors in cyclic triaxial tests.

4.5 Conclusions

The soft clay, as a viscoelastic plasticity soil, is widely distributed in Shanghai. The dynamic characteristics of the saturated soft clay under the cyclic vibration loading which determine the safe operation of subway have been one of research interests particularly in the last decade. In this chapter, 12 cyclic triaxial tests have been carried out to investigate the dynamic characteristics of the silty clay before and after freezing-thawing, including the dynamic modulus, the damping ratio, the dynamic strain and the excess pore water pressure, etc. The main conclusions are drawn as follows.

(1) The damping ratio of soil represents the capability of samples to absorb energy under cyclic loading. The larger the damping ratio is; the more energy is absorbed by samples. The energy transformed into soil partly leads the plastic deformation of samples and partly increases the excess pore pressure under the cyclic loading. The energy can be represented by the areas enclosed by the load and unload hysteretic curves, which are well fitted by the geometric hyperbolic functions.

Fig. 4.12 The axial strain of freezing-thawing and undisturbed samples

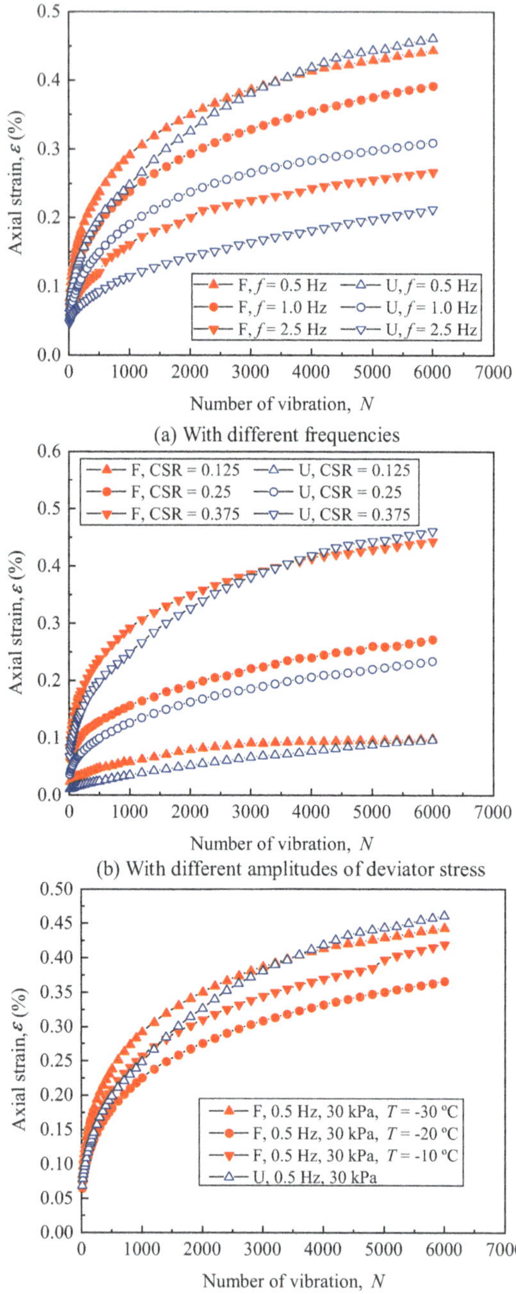

(c) With different freezing temperatures

(2) The frequency and the CSR have significant effects on the dynamic behaviors of the saturated soft clay. The lower the frequency and the larger the CSR are, the bigger the dynamic modulus degradation is. The modulus decreases significantly when the strain is small. The dynamic modulus increased after freezing-thawing under the same cyclic loading condition. The temperature of −10 °C can be considered as an inflected point of freezing temperature that determines the dynamic characteristics of freezing-thawing clay. When the temperature drops to −10 °C in the freezing process, the disturbance from ice extrusion to clay structures is the most serious after thawing. The modulus of the freezing-thawing sample with −30 °C is 12.2% larger than that of the undisturbed sample under the same conditions.
(3) With the increasing of number of vibration, the damping ratio curves show a downward trend. The decline of damping ratio before and after triaxial tests increases with the freezing temperature dropping. The lower the temperature is, the greater the pore volumes are expanded and more room is obtained for deformation of soil samples after thawing. For the freezing-thawing sample with temperature of −30 °C, the damping ratio drops about 50% after cyclic triaxial tests.
(4) With the increasing of number of vibration, the increment of deformation and the excess pore water pressure within samples under the cyclic loading become smaller and smaller. The lower the frequency and the larger the amplitude of deviator stress are, the larger the axial strain and the higher excess pore pressure develop. But the rates decrease gradually with the number of vibration increasing. The excess pore water pressure and axial strain increase after freezing-thawing. The lower the freezing temperature is, the higher the excess pore water pressure develops. The axial strains of samples with CSRs of 0.25 and 0.125 increase 16% and 5% after freezing-thawing, respectively.

References

Christ, M., Kim, Y. C., & Park, J. B. (2009). The influence of temperature and cycles on acoustic and mechanical properties of frozen soils. *KSCE Journal of Civil Engineering, 13*(3), 153–159.

Cui, Z. D., He, P. P., & Yang, W. H. (2014). Mechanical properties of a silty clay subjected to freezing-thawing. *Cold Regions Science and Technology, 98,* 26–34.

Cui, Z. D., & Zhang, Z. L. (2015). Comparison of dynamic characteristics of the silty clay before and after freezing and thawing under the subway vibration loading. *Cold Regions Science and Technology, 119,* 29–36.

Hanumantharao, C., & Ramana, G. V. (2008). Dynamic soil properties for microzonation of Delhi, India. *Journal of Earth System Science, 117*(2), 719–730.

Hardin, B. O., & Drnevich, V. P. (1972a). Shear modulus and damping in soils. *Journal of the Soil Mechanics and Foundations Division, 98*(7), 667–692.

Hardin, B. O., & Drnevich, V. P. (1972b). Shear modulus and damping in soils: Measurement and parameter effects (terzaghi leture). *Journal of the Soil Mechanics and Foundations Division, 98*(6), 603–624.

Idriss, I. M., Dobry, R., & Sing, R. D. (1978). Nonlinear behavior of soft clays during cyclic loading. *Journal of Geotechnical and Geoenvironmental Engineering, 104*(GT12), 1427–1447.

Kang, M., & Lee, J. S. (2015). Evaluation of the freezing–thawing effect in sand–silt mixtures using elastic waves and electrical resistivity. *Cold Regions Science and Technology, 113,* 1–11.

Kokusho, T. (1980). Cyclic triaxial test of dynamic soil properties for wide strain range. *Soils and Foundations, 20*(2), 45–60.

Konrad, J. M. (1989). Physical processes during freeze-thaw cycles in clayey silts. *Cold Regions Science and Technology, 16*(3), 291–303.

Lee, C. J., & Sheu, S. F. (2007). The stiffness degradation and damping ratio evolution of Taipei Silty Clay under cyclic straining. *Soil Dynamics and Earthquake Engineering, 27*(8), 730–740.

Li, S., & Wang, J. (2013). Constitutive model of saturated soft clay with cyclic loads under unconsolidated undrained condition. *Transactions of Tianjin University, 19,* 260–266.

Ling, X., Li, Q., Wang, L., et al. (2013). Stiffness and damping radio evolution of frozen clays under long-term low-level repeated cyclic loading: Experimental evidence and evolution model. *Cold Regions Science and Technology, 86,* 45–54.

Ling, X., Zhu, Z., Zhang, F., et al. (2009). Dynamic elastic modulus for frozen soil from the embankment on Beiluhe Basin along the Qinghai-Tibet railway. *Cold Regions Science and Technology, 57*(1), 7–12.

Liu, G., & Kai, Q. (2010). *Dynamic property of mucky soil of subgrade under cyclic load.* Shanghai: Soil Dynamics and Earthquake Engineering.

Luan, M., Liu, G., Wang, Z., et al. (2010). Stiffness degradation of undisturbed saturated soft clay in the yangtze estuary under complex stress conditions. *China Ocean Engineering, 24*(3), 523–538.

Martin, P. P., & Seed, H. B. (1982). One-dimensional dynamic ground response analyses. *Journal of the Geotechnical Engineering Division, 108*(7), 935–952.

Masing, G. (1926). Self stretching and hardening for brass. In *Proceedings of the Second International Congress for Applied Mechanics*. Zurich, Switzerland.

Mortezaie, A. R., & Vucetic, M. (2013). Effect of frequency and vertical stress on cyclic degradation and pore water pressure in clay in the ngi simple shear device. *Journal of Geotechnical and Geoenvironmental Engineering, 139*(10), 1727–1737.

Okur, D. V., & Ansal, A. (2007). Stiffness degradation of natural fine grained soils during cyclic loading. *Soil Dynamics and Earthquake Engineering, 27*(9), 843–854.

Puzrin, A., Frydman, S., & Talesnick, M. (1995). Normalized nondegrading behavior of soft clay under cyclic simple shear loading. *Journal of Geotechnical Engineering, 121*(12), 836–843.

Seed, H. B., & Idriss, I. M. (1970). *Soil moduli and damping factors for dynamic response analyses.* Berkeley, California: Earthquake Engineering Research Center, University of California.

Seed, H. B., Wong, R. T., Idriss, I. M., et al. (1986). Moduli and damping factors for dynamic analyses of cohesionless soils. *Journal of Geotechnical Engineering, 112*(11), 1016–1032.

Tang, Y. Q., Cui, Z. D., Zhang, X., et al. (2008). Dynamic response and pore pressure model of the saturated soft clay around the tunnel under vibration loading of Shanghai subway. *Engineering Geology, 98*(3), 126–132.

Vucetic, M., & Dobry, R. (1988). Degradation of marine clays under cyclic loading. *Journal of Geotechnical Engineering, 114*(2), 133–149.

Vucetic, M., & Mortezaie, A. (2015). Cyclic secant shear modulus versus pore water pressure in sands at small cyclic strains. *Soil Dynamics and Earthquake Engineering, 70,* 60–72.

Wang, T., Liu, Y., Yan, H., et al. (2015). An experimental study on the mechanical properties of silty soils under repeated freeze–thaw cycles. *Cold Regions Science and Technology, 112,* 51–65.

Zhang, Y., & Hulsey, J. L. (2015). Temperature and freeze-thaw effects on dynamic properties of fine-grained soils. *Journal of Cold Regions Engineering, 29*(2), 04014012.

Zhang, Z.L., & Cui, Z.D. (2018). Effect of freezing-thawing on dynamic characteristics of the silty clay under K0-consolidated condition. *Cold Regions Science and Technology, 146,* 32–42.

Zhou, J., & Tang, Y. (2015). Artificial ground freezing of fully saturated mucky clay: Thawing problem by centrifuge modeling. *Cold Regions Science and Technology, 117,* 1–11.

Chapter 5 Microscopic Pore Structures of Silty Clay Before and After Freezing-Thawing by SEM

5.1 Introduction

With the development of urban rail transposition in Shanghai, the long-term deformation of soft clay under the subway vibration loading has recently drawn more attention by researchers and engineers. The soft clay, with multiscale heterogeneity of structure, is characterized by high water content, high compressibility, high sensitivity, low permeability and low bearing capacity (Cui and Zhang 2015). Generally, the soft clay mainly consists of two parts, skeleton particles (original minerals, clay minerals, soluble salts, etc.) and pore water. However, the complexity of soft clay is not only influenced by the mineral composition, but also by the pore water between aggregate structures (Zhang and Li 2010; Wei et al. 2013). With an increase in the water content of soil, the soil structure changes from a good level of integration to flocculation such that the particles connect to each other mainly by edges and angularities (Liu et al. 2010). These characteristics of soft clay with water-soil interactions have brought great challenges to the construction engineers, besides new scientific problems to researchers.

To guarantee the safety in the construction of subway tunnel in soft soil areas, the artificial freezing method has been widely utilized in the by-passes of the subway tunnel due to its unique advantage as both the temporary support and the water barrier. Tang et al. (2012) analyzed the changing of the pores in soft clay before and after the freezing-thawing, and found that the average pore diameters become larger, due to the conversion between the pore water and ice crystals during freezing-thawing. Chamberlain and Gow (1979) found a reduction in void ratio and an increase in vertical permeability with four fine-grained soils before and after freezing-thawing, which attributed to the formation of polygonal shrinkage cracks. With the development of the freezing process, the ice crystal aggregates gradually become the significant factor of soil particles. The denser the microstructure is, the stronger the ice crystals aggregates, and soil exhibits higher mechanical strength (Cui et al. 2014; Liu and Zhang 2014). Moreover, large thawing settlement appeared in the centrifuge

Z.-D. Cui et al., *Dynamics of Freezing-Thawing Soil around Subway Shield Tunnels*,
https://doi.org/10.1007/978-981-15-4342-5_5

model test, with great improvement in permeability after artificial ground freezing and thawing (Zhou and Tang 2015). Irrespective of how much deformation of the clay was, the irreversible mechanical changes occurred in the microstructure of the mineral particles, manifested by deformations, cracks and disintegration of mineral particles (Ewa 2015).

The deformation, the cumulative results of soil structure change in nature, including the movement and arrangement of particles and particle aggregates, happened to soil samples under load (Vasseur et al. 1995; Cui et al. 2016). Depending on the size of particles and pores, the structure of soil can be classified. The soil structure is divided into macrostructure and microstructure by the commonly-used classification (Liu et al. 2010). The microstructure (<100 μm) refers to the crystal structure and mineral composition which can be measured by SEM, mercury intrusion porosimetry (MIP), transmission electron microscopy (TEM), X-ray diffraction (XRD), computerized X-ray tomography (CT) and so forth (Ninjgarav et al. 2007; Wei et al. 2013). With the development of the test technologies, the microstructures of soil can be analyzed more and more accurately. In addition, above two or more methods can be combined through data-fusion to analyze the organization and orientation of the particles and pores, providing accuracy scopes for researching the 2D to 3D registration of soils structure (De Boever et al. 2015; Hemes et al. 2015).

In terms of testing technologies of porous structures and quantitative analysis they provide, it is certain that there are plenty of useful researching results of the soil microstructure. For instance, the Otsu method of picking up inter-aggregate pores of soil from the SEM images depending on a certain grayscale value, which named threshold, was put forward by Cui and Jia (2013). Dathe et al. (2001) pointed out a reduction of fractal dimension values with the magnification of SEM images by different fractal models increasing. Liu et al. (2005) researched effects of the image magnification on other quantitative parameters of soil SEM images and adopted a "window method" (box counting) to study the relationship between the size proportion and the quantitative analysis result of SEM images. However, it was ascertained by Bird et al. (2006) that SEM images processing cannot display the fractal behavior of soils completely, especially at the range of scales accessible to analysis. In addition, the depolarization mechanism with particle orientation analyses under loading was observed by Hattab and Fleureau (2010, 2011) through the isotropic triaxial tests. The variation of pore microstructures can be investigated with the probability distribution index, even though it was merely applied to SEM images with an appropriate magnification, the pores area of which was greater than 50 pixels (Liu et al. 2011).

5.2 The SEM Test

The SEM tests were carried out after the cyclic triaxial tests finished (Zhang and Cui 2017). The SEM equipment used in this research was Hitachi SU1510. The high resolution SEM allows detailed imaging and analysis of soil microstructures with resolution of 3 nm scale. The accelerating voltage used in the tests was 15.0 kV, and the magnification range was between 1000 and 5000. SEM observations must be carried out on completely dry samples. There are several ways to remove the water from a clay sample while keeping the fabric of the material invariant. The simplest way was "freeze-drying", where the samples were generally frozen in liquid nitrogen and freeze-dried afterward (Shi et al. 1999). Consequently, the ice was sublimated excluding passing through liquid state, avoiding parasitic suction effect. In this research, the "freeze-drying" method was applied. In order to ensure the SEM images reveal the true microstructure of the clay, the preparation process of clay samples was as follows.

(1) Turn off each sample from the triaxial tests apparatus and remove the rubber membrane carefully, maintaining the original condition of each sample.
(2) Incise a wedge notch along the height of each sample with a sharp blade. Cut the middle part of each sample into strips of approximately 10 × 10 × 40 mm.
(3) Put the strips in liquid nitrogen to be frozen instantaneously. Subsequently put them into the vacuum freeze-drying machine over 24 h to be dried excluding microstructures disturbed.
(4) Separate the dried strips into certain sizes along the observing height direction to guarantee the uniformity and presentation. As the vertical cross-section was the observed microstructure surface of samples, choose the middle part of the strip with surface as smooth as possible.
(5) Fix the samples on the conductive base with white latex and spray a thin gold layer, for the electrical conductivity of clay is poor. Operate the equipment by software, the SEM images with different magnifications were obtained.

5.3 Test Results

In order to research the variations of microstructure of silty clay before and after freezing-thawing, the SEM tests were conducted with samples subjected to cyclic loading. SEM images with magnifications of 1000×, 2000×, 3000× and 5000× were obtained, summarized in Table 5.1. Part of the SEM images with magnification of 3000× and size of 1280 × 960 pixel are shown in Fig. 5.1. In order to clearly see the microscopic pore and particle structures of the samples, the gray scale value of each pixel in the SEM images are obtained as a height to reconstruct the 3D surfaces of the sample sections. As can be seen from the images, the soil particles show an irregular flake structure, leaving a rough surface with a three-dimensional sense. The adjacent particles may gather into particle aggregates, resulting in amorphous pores. In order to study the variations of the microstructure of the soil before and after freezing-thawing under cyclic loadings, the SEM images processing was conducted for the quantitative analysis.

Table 5.1 Schemes of undrained cyclic triaxial tests

No.	Soils	Freezing temperature (°C)	Confining pressure (kPa)	Axial stress (kPa)	Deviator stress (kPa)	f (Hz)	CSR
01	F	−30	240	300	30	0.5	0.375
02	F	−30	240	300	30	1.0	0.375
03	F	−30	240	300	30	2.5	0.375
04	F	−30	240	300	20	0.5	0.25
05	F	−30	240	300	10	0.5	0.125
06	F	−20	240	300	30	0.5	0.375
07	F	−10	240	300	30	0.5	0.375
08	F	−30	–	–	–	–	–
09	U	–	240	300	30	0.5	0.375
10	U	–	240	300	30	1.0	0.375
11	U	–	240	300	30	2.5	0.375
12	U	–	240	300	20	0.5	0.25
13	U	–	240	300	10	0.5	0.125
14	U	–	–	–	–	–	–

5.3.1 Quantitative Analysis of Test Results

The soft clay can be divided into two parts, the soil aggregates and the pores including inter-aggregates pores and intra-aggregates pores. For the convenience of analysis, the SEM image is first converted to a binary image, with the values of 0 (white pixels) and 1 (black pixels) representing the soil aggregates and the pores, respectively. If the grayscale value of a certain pixel is smaller than or equals to that of the determined threshold, it is identified as the pore; otherwise, it is identified as the soil aggregate. However, it is difficult to accurately determine what grayscale value represents the soil aggregates or pores in some grayscale images. Take the SEM image of the freezing-thawing sample (No. 08) with magnifications of 3000× as an example. Figure 5.2 shows the pore structures with thresholds being 0.1, 0.2, 0.3 and 0.4, respectively. The larger the adopted threshold is, the larger the void ratio will be. With an appropriate threshold, the soil aggregates and pores can be differentiated truly and reasonably. However, the threshold in many literatures is determined by the authors' judgment on the interface between pores and soil aggregates for simplification, which may be an argument to aid the calibration of measured porosity.

To obtain the threshold for a certain SEM image, the Otsu method can be used (Otsu, 1975). That is, suppose t as the dividing threshold between the foreground and the background, ω_0 and ω_1 being the ratio of the pixels in the foreground and in the background to all the pixels, respectively, and u_0 and u_1 being the mean grayscale value of the pixels in the foreground and in the background, respectively, the mean grayscale value of the whole image is $u = \omega_0 \times u_0 + \omega_1 \times u_1$. When the variance

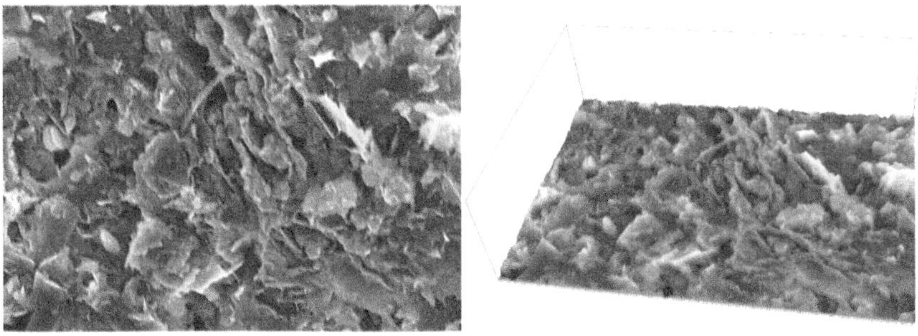

(a) No. 01, 0.5 Hz, 30 kPa, F (-30 ºC)

(b) No. 09, 0.5 Hz, 30 kPa, U

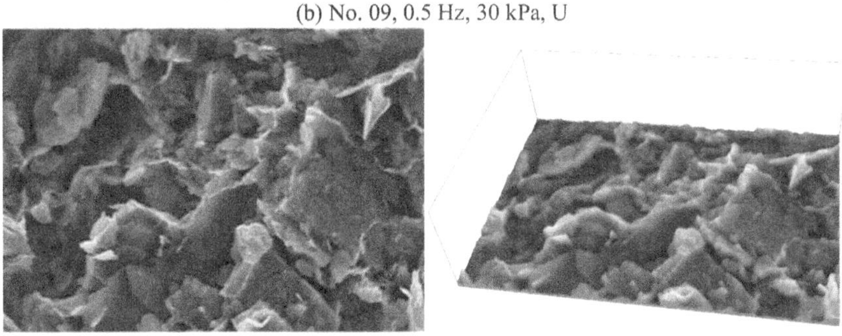

(c) No. 08, F (-30 ºC)

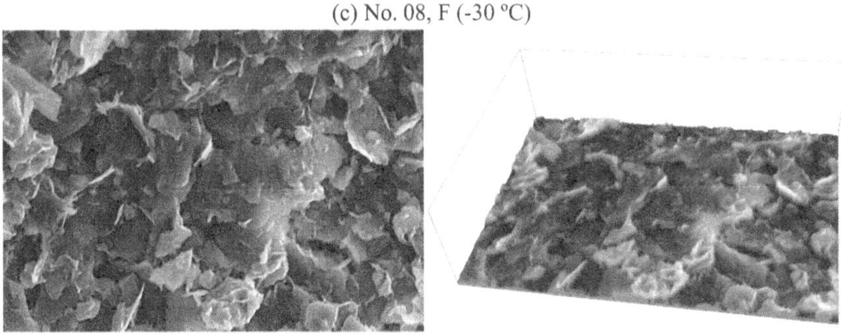

(d) No. 14, U

Fig. 5.1 SEM images of each sample

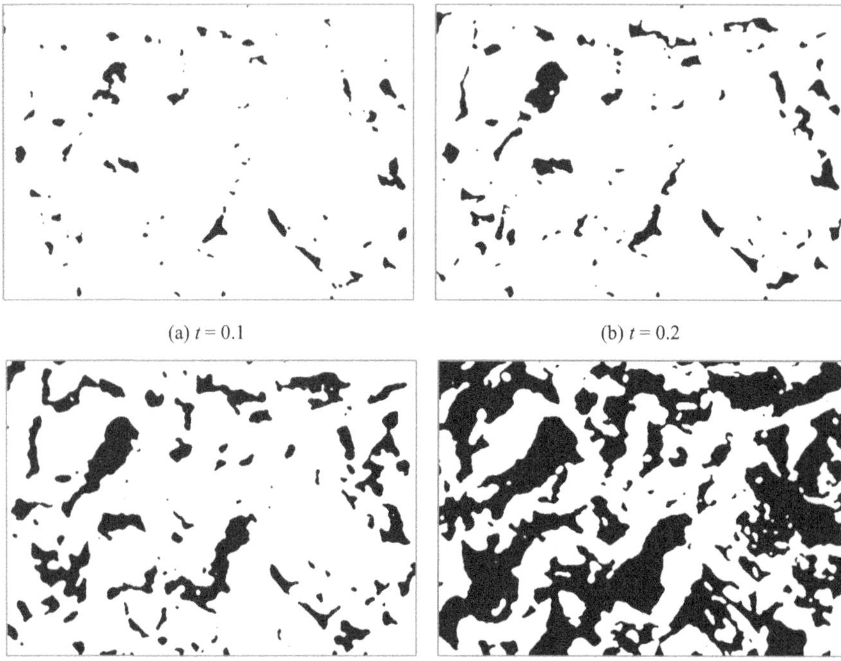

(a) $t = 0.1$ (b) $t = 0.2$

(c) $t = 0.3$ (d) $t = 0.4$

Fig. 5.2 The pore structures of freezing-thawing sample (No. 08) with different thresholds

$g = \omega_0(u_0 - u)^2 + \omega_1(u_1 - u)^2 = \omega_0 \times \omega_1 \times (u_0 - u_1)^2$ reaches the maximum, t is the best dividing threshold. Variations of the porosity n of each SEM image with magnification of 3000× to different thresholds calculated by the Otsu method are summarized in Table 5.2. However, it is not accurate enough to calculate the porosity of soils by the Otsu method. The porosity obtained from the two-dimensional image processing was seriously affected by the selected threshold while the vertical and horizontal scale of the image should be adjusted based on the experience for the three-dimensional processing (Wang et al. 2004). For example, the real porosity of the silty clay is about 0.5 calculated from the void ratio. The minimum and maximum errors of the porosity by the Otsu method for the silty clay samples are 20.3 and 50.32%, respectively. So the threshold in the Otsu method needs to be corrected.

Cui and Jia (2013) compared the curves of soil filling rate ($n' = 1 - n$) versus different thresholds based on the Otsu method and found that the corresponding threshold to the filling rate can be back calculated by the polynomial fitting function between the filling rate and the threshold. After repeating this process many times, a relational fitting curve of the real filling rate versus the threshold of SEM images is generated, as shown in Fig. 5.3. The corresponding threshold to the real filling rate can be back calculated by the fitting function between n' and t. The real filling rate can be estimated by adding 15% to the calculated filling rate of the particle with the

Table 5.2 The porosity of the silty clay samples calculated with different thresholds

No.	0.1	0.2	0.3	0.4	0.5	0.6	0.7	0.8	0.9	Otsu	Otsu threshold	Otsu error (%)
01	0.0171	0.1126	0.3540	0.5900	0.7635	0.8877	0.9525	0.9833	0.9956	0.6295	127	25.9
02	0.0285	0.2250	0.5232	0.7584	0.8942	0.9639	0.9927	0.9995	1.0000	0.6337	116	26.74
03	0.0001	0.0590	0.3144	0.6812	0.8404	0.9304	0.9702	0.9870	0.9940	0.7516	129	50.32
04	0.0198	0.1696	0.4174	0.6638	0.8221	0.9201	0.9730	0.9966	1.0000	0.6398	119	27.96
05	0.0252	0.1961	0.4470	0.6950	0.8541	0.9380	0.9827	0.9973	0.9999	0.6184	121	23.68
06	0.0166	0.1679	0.4300	0.6523	0.7930	0.8921	0.9538	0.9851	0.9980	0.6725	127	34.5
07	0.0362	0.2886	0.5560	0.7439	0.8694	0.9519	0.9870	0.9976	0.9997	0.6613	125	32.26
08	0.0212	0.2268	0.5268	0.7367	0.8606	0.9414	0.9794	0.9970	0.9999	0.6928	123	38.56
09	0.0946	0.3031	0.5278	0.7298	0.8640	0.9467	0.9817	0.9954	0.9998	0.6017	120	20.34
10	0.0328	0.2501	0.5653	0.7884	0.9170	0.9771	0.9932	0.9994	1.0000	0.6480	118	29.6
11	0.0450	0.3026	0.6149	0.8421	0.9360	0.9833	0.9980	1.0000	1.0000	0.6378	115	27.56
12	0.0212	0.1475	0.3868	0.6551	0.8325	0.9327	0.9766	0.9947	0.9997	0.6180	125	23.6
13	0.0328	0.2339	0.5140	0.7404	0.8879	0.9662	0.9926	0.9993	1.0000	0.6084	119	21.68
14	0.0137	0.2044	0.4802	0.6785	0.8116	0.9161	0.9725	0.9944	0.9994	0.6597	126	31.94

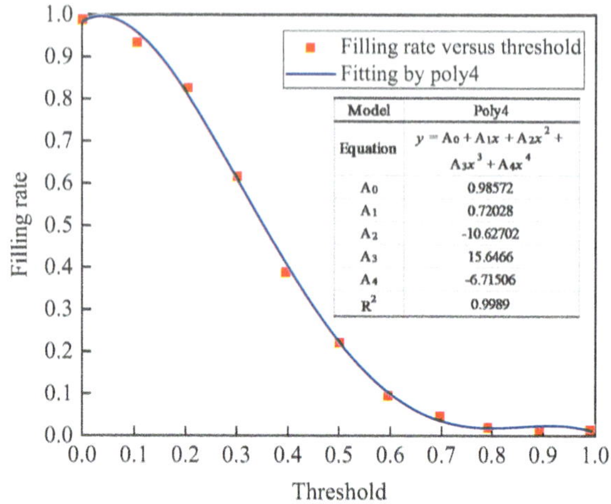

Fig. 5.3 Variations of filling rate versus threshold of SEM image (No. 08)

threshold. For the silty clay of layer No. 5 in Shanghai, the porosity is estimated by reducing about 30% to the calculated porosity with the Otsu method. The selected threshold value is in the range [0.32, 0.36], which corresponds to the value between the range [81, 90] for the real grayscale. In this chapter, the threshold was selected as [85] for convenience, with porosity error within 5%.

5.3.2 Effects of Magnification on Porosity from SEM Images

For the clay SEM images, not only the threshold but also the magnification are essential to display the microstructure of silty clay realistically. The magnification, the ratio of resolution of human eyes to that of SEM equipment, is based on limitation of resolution of human eyes on the screen. The larger the magnification is, the smaller range with more retails can be observed on the SEM images. However, the range of clay SEM images is expected to reveal the original microstructures with the porosity as accurate as possible. If the magnification is extremely small, the boundary between the particle aggregates and pores may be ignored in the process of binary. If too large, the results may be lack of representation. Therefore, effects of the magnification on the interpretation of clay SEM images should be discussed and an appropriate magnification of the SEM images should be selected to study the microstructure of soils quantitatively. Figure 5.4 shows the binary images of undisturbed sample of No. 09 with the magnifications of 1000×, 2000×, 3000×, and 5000×, respectively.

According to the selected threshold, the binaryzation of images with different magnifications were processed and the porosity of the silty clay samples from the binary images were calculated. Figure 5.5 presents the distribution of clay porosity from the SEM images with different magnifications. It can be seen that all the mean porosity of silty clay with the same magnifications is close to the real porosity of silty

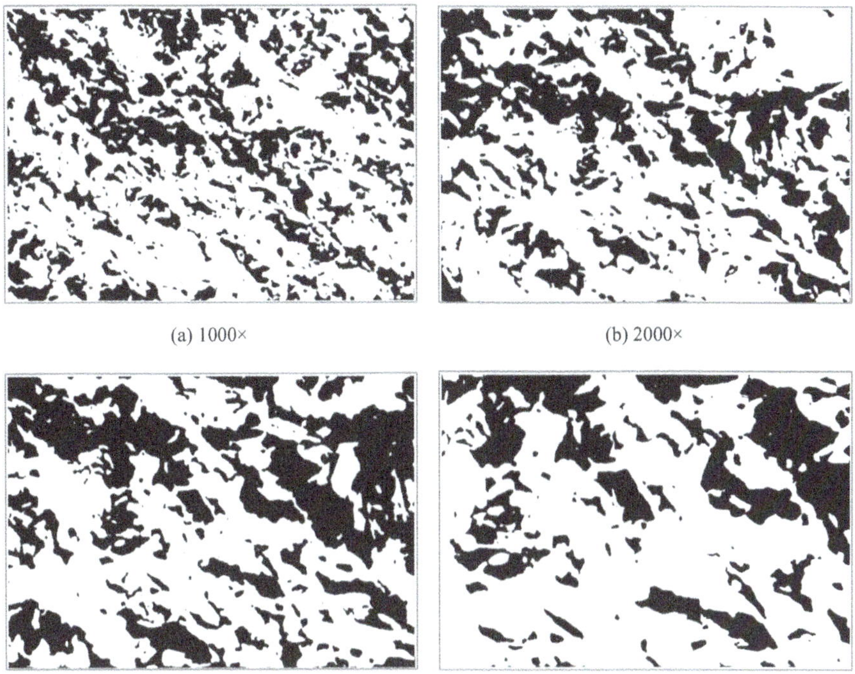

(a) 1000×

(b) 2000×

(c) 3000×

(d) 5000×

Fig. 5.4 The binary images of undisturbed sample with frequency of 0.5 Hz and deviator stress of 30 kPa (No. 09)

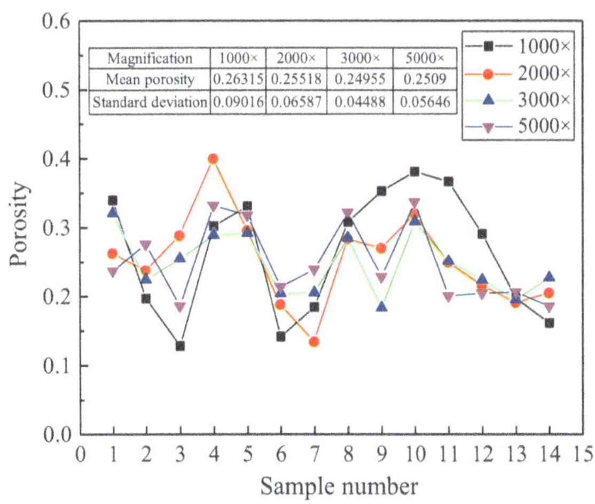

Fig. 5.5 Distribution of porosity from the SEM images with different magnifications

clay samples by reducing 30%. The results contributed to the improvement of the Otsu method by Cui and Jia (2013). The porosity of each silty clay sample through the SEM images with different magnifications is summarized in Table 5.3. The mean values and the standard deviation of the porosity with the same magnification are compared and analyzed. The magnification of 3000× is relatively accurate and shows less fluctuation of the porosity data for both freezing-thawing and undisturbed samples comparing with other magnifications, even though they all meet the requirements and the results are close to each other. As a result, the magnifications from 1000× up to 5000× are all suitable for observation of the clay microstructures. The effect of magnification from 1000× up to 5000× is not seriously significant in the porosity analysis of SEM images for the clay SEM tests, while other magnifications still need more research. In this chapter, the magnification of the silty clay SEM images was selected as 3000× for further quantitative analysis.

5.3.3 Further Analysis of Microscopic Pore Structures

The SEM tests of silty clay before and after freezing-thawing under cyclic loading were conducted to study the microstructures of silty clay. The approach consists of analyzing the microscopic pore structures of the soil using binary images converted from SEM images and extracting the quantitative parameters, including the pore number, the mean diameter, the pore shape coefficient, the porosity, the frequency distribution of pores orientation, the pore anisotropy, and the fractal dimension etc.

5.3.3.1 Pore Shape Coefficient

The pore shape coefficient K, as one of the most significant characteristic parameters of the clay pores, is used to describe the variation of the microscopic pore structures. As Eq. (5.1) showed, it refers to the ratio of the equivalent perimeter to the real perimeter of the pore, whose value is between 0 and 1. For the shapes with the same area, the smaller the perimeter is, the smoother the pore shapes and the larger the K values are.

$$K = \frac{2\pi\sqrt{A/\pi}}{P} \tag{5.1}$$

where A is the real area of each pore and P is its perimeter. In addition, the shape characteristics can be expressed by the circularity c of the pores, which is the square of the pore shape coefficient K. Similar to the pore shape coefficient, the circularity can be used to assess the level of the pore shape close to the circle. The smaller the circularity values, the slenderer the pore shapes.

The statistical distribution of the pore shape coefficient of the silty clay samples at various triaxial cyclic tests is presented in Fig. 5.6. It can be seen that the trends of the

Table 5.3 The porosity of the silty clay samples through the SEM images with different magnifications

No.	Soils	f (Hz)	CSR	1000×	2000×	3000×	5000×
01	F (−30 °C)	0.5	0.375	0.184693801	0.1363165	0.207375	0.241048474
02	F (−30 °C)	1.0	0.375	0.30815868	0.2856212	0.287225	0.323832463
03	F (−30 °C)	2.5	0.375	0.353012092	0.2724328	0.185437	0.23036299
04	F (−30 °C)	0.5	0.25	0.381296372	0.3220096	0.310536	0.339124281
05	F (−30 °C)	0.5	0.125	0.366735301	0.2515491	0.253221	0.202537023
06	F (−20 °C)	0.5	0.375	0.291138325	0.2182718	0.225876	0.206524411
07	F (−10 °C)	0.5	0.375	0.197152316	0.1934326	0.197989	0.20822722
08	F (−30 °C)	–	–	0.161542143	0.2074019	0.229334	0.187797907
09	U	0.5	0.375	0.339853824	0.2648979	0.322548	0.238562525
10	U	1.0	0.375	0.196937204	0.2399756	0.225981	0.277839793
11	U	2.5	0.375	0.128440567	0.2905297	0.257041	0.187793982
12	U	0.5	0.25	0.302381952	0.4017527	0.290928	0.333712327
13	U	0.5	0.125	0.330980891	0.2980731	0.294011	0.319710405
14	U	–	–	0.141824145	0.1902909	0.206161	0.215482707

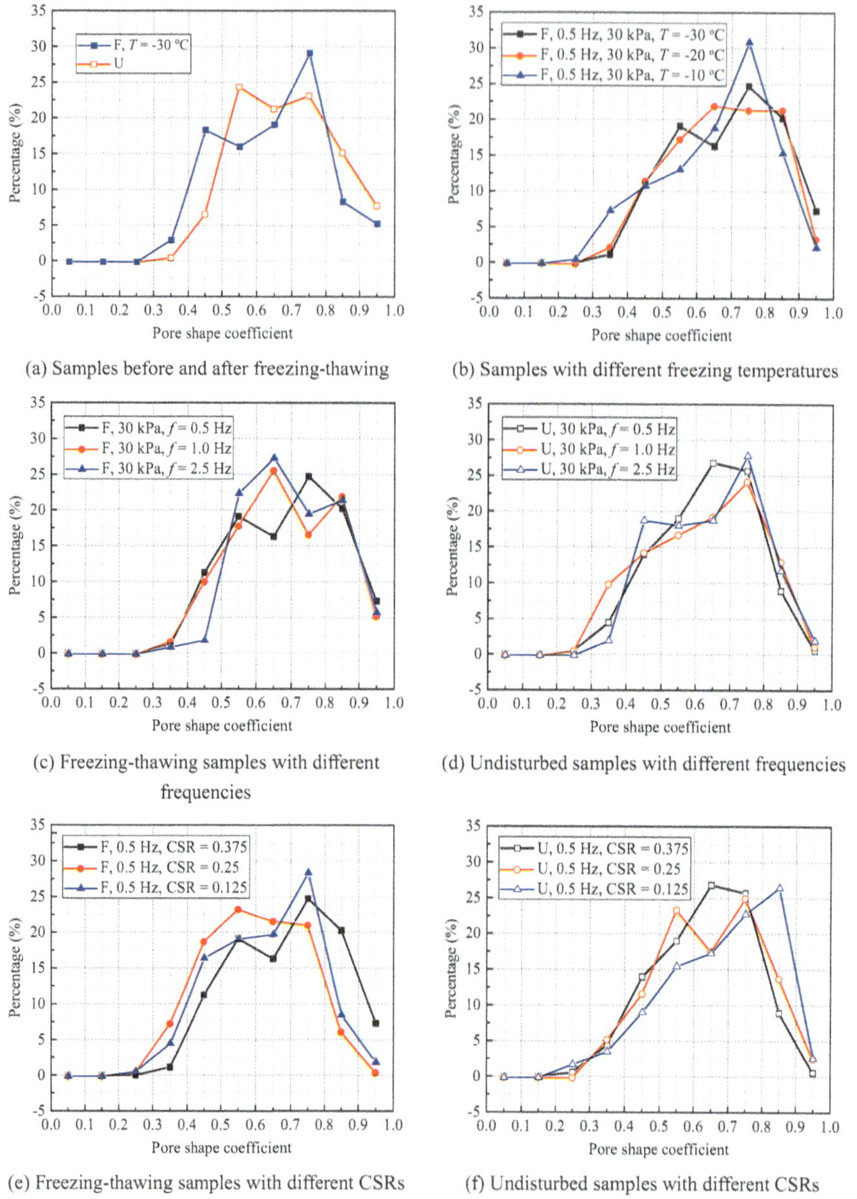

(a) Samples before and after freezing-thawing

(b) Samples with different freezing temperatures

(c) Freezing-thawing samples with different frequencies

(d) Undisturbed samples with different frequencies

(e) Freezing-thawing samples with different CSRs

(f) Undisturbed samples with different CSRs

Fig. 5.6 The distribution curves of pores shape coefficient distribution are consistent with the overall curves. Most of the pore shape coeffecients are distributed in [0.45, 0.85]. Comparing the mean pore shape coefficient with the values summarized in Table 5.4, it indicates that some of the pores are smoothed after freezing-thawing. The mean pore shape coefficient of the freezing-thawing clay without loading is improved 0.53% than that of the undisturbed clay. The coefficient of the freezing-thawing clay under cyclic loading with frequency of 0.5 Hz and CSR of 0.375 is improved 0.76% than that of the undisturbed clay. Therefore, the freezing-thawing process has little effect on the pore shape coefficient of the silty clay. Actually, the improvement of pore shape coefficient derived from the variation of the soil particle arrangement is not conducive to the utility of soil friction. After freezing-thawing, the shear strength of the silty clay is reduced. It is considered that when the temperature reaches below the freezing point, the water in the saturated soil is converted into ice. The volume of soil expands and gradually the pores are full of ice among the soil particles, resulting in the sliding of the particles and the deformation of the pore shapes (Othman and Benson 1993). After thawing, the ice in the soil changes back to water, which reduces the contacts among the soil particles. As a result, the shear strength of the soil is reduced, coupled with the decrease of the cohesive force, the bonding strength, and the bite force among the particles.

5.3.3.2 Pore Orientation Distribution

For the soft clay under the cyclic loadings, most of the soil particles are compacted and aggregated with the variation of pore orientation. It is valuable to research the orientation distribution of soil, coupled to the observation surface of the SEM tests being vertical and the identification of the preferential orientations of the pores with respect to the horizontal plane. In this chapter, the frequency distribution of the pore orientation was analyzed to assess the distribution of the pores within the samples. Since the range of pore orientation angles is [0°, 180°], for the sake of the integrity of statistical regularities, 180° is added to the data to ensure the orientation angle being centrosymmetric. The orientation is divided into 24 groups, 15° for each group. Accordingly, the statistical frequency distribution of the pore orientation is calculated as Eq. (5.2) showed.

$$F_i(\alpha) = \frac{m_i}{M} \times 100\% \tag{5.2}$$

where m_i is the number of pores in each group and M is the total number of pores in all groups. In addition, the frequency of the pores can be substituted with the number of pores. Figure 5.7 presents the wind rose diagrams of the pore orientation distribution with the number of pores. The frequency orientation distribution of pores is random, which signifies that the distribution of the microstructure of the clay is homogeneous. Group with more pores exbhits higher orientation. In order to analyze the homogeneity and random of the pore orientation, the anisotropy I_n is defined as Eq. (5.3) showed.

Table 5.4 Summary of the microstructure parameters of silty clay

No.	Soils	f (Hz)	CSR	Axial strain increment (%)	Pore pressure increment (kPa)	Pore number	Porosity (%)	Mean pore shape coefficient	Anisotropy (%)	Fractal dimension
01	F (−30 °C)	0.5	0.375	0.3574	11.1588	179	20.7375	0.85181	24	1.22756
02	F (−30 °C)	1.0	0.375	0.3067	11.2382	168	28.7225	0.84077	36.36	1.202
03	F (−30 °C)	2.5	0.375	0.2004	8.0906	103	18.5437	0.85431	50	1.1536
04	F (−30 °C)	0.5	0.25	0.2295	5.5298	176	31.0536	0.83252	25	1.22974
05	F (−30 °C)	0.5	0.125	0.0678	5.2852	152	25.3221	0.84818	40.91	1.24072
06	F (−30 °C)	0.5	0.375	0.294	11.5963	173	22.5876	0.85375	35.71	1.20854
07	F (−30 °C)	0.5	0.375	0.345	12.7303	176	19.7989	0.88046	26.67	1.26328
08	F (−30 °C)	–	–	–	–	131	22.9334	0.86305	35.29	1.21238
09	U	0.5	0.375	0.4254	8.5693	180	32.2548	0.84537	44.44	1.23884
10	U	1.0	0.375	0.2572	8.4273	161	22.5981	0.83325	63.64	1.23664
11	U	2.5	0.375	0.1567	5.9226	144	25.7041	0.85635	22.73	1.18302
12	U	0.5	0.25	0.1981	8.6205	188	29.0928	0.83550	9.09	1.24302
13	U	0.5	0.125	0.0857	4.0853	111	29.4011	0.83593	13.33	1.26208
14	U	–	–	–	–	164	20.6161	0.85848	30	1.2153

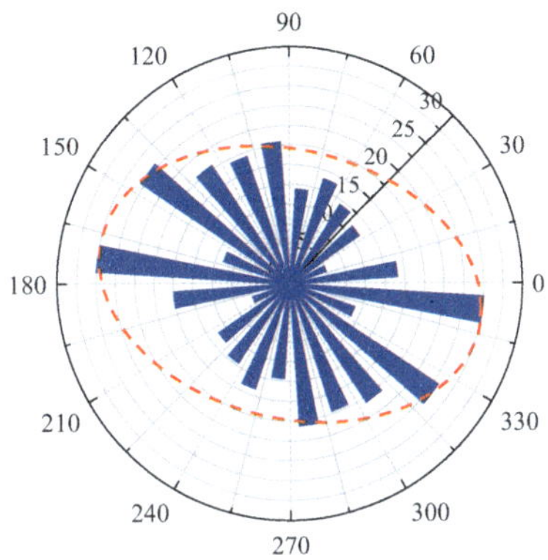

(a) Freezing-thawing samples without loading (No. 08)

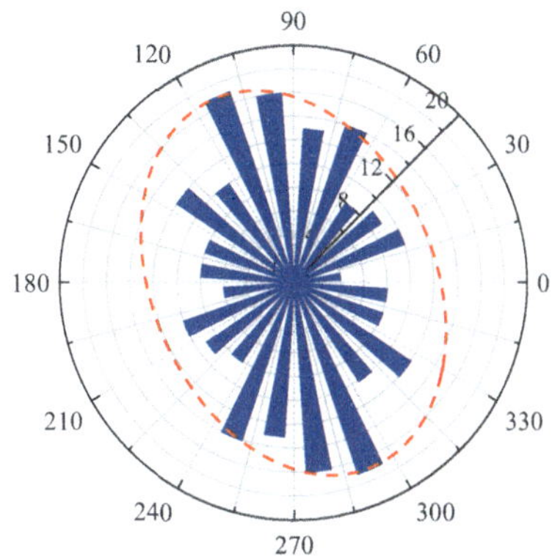

(b) Freezing-thawing samples with frequency of 0.5 Hz and CSR of 0.375 (No. 01)

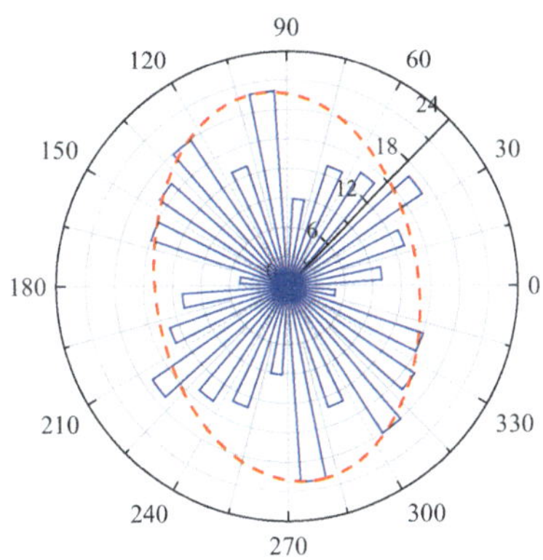

(c) Undisturbed samples without loading (No. 14)

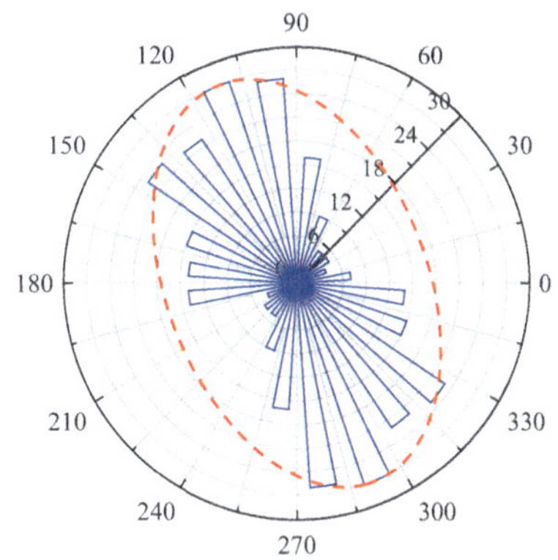

(d) Undisturbed samples with frequency of 0.5 Hz and CSR of 0.375 (No. 09)

Fig. 5.7 Distribution of pore orientation angles

$$I_n = \frac{R - r}{R} \times 100\% \tag{5.3}$$

where R is the long axis of the ellipse derived from the orientation radius in the wind rose diagram; r is the short axis of the ellipse. The results of the anisotropy are summarized in Table 5.4. The larger the anisotropy values, the more consistent the distribution of the pores.

It can be seen from Fig. 5.7 that the wind rose diagrams are centrosymmetric. The pores orientation of samples without loadings are arranged horizontally, while the pores of samples under cyclic loadings are arranged vertically from the distribution of the pore orientation. On one hand, it means that the microstructures are indeed disturbed by the cylic loadings. On the other hand, it is an indication that orientation frequency distribution of a SEM image is suitable for analyzing the variation of the SEM images. In addition, the mean anisotropy value of the freezing-thawing clay are 12% larger than that of the undisturbed clay, implying an effect of the freezing-thawing on the clay microstructures.

5.3.3.3 Pore Size Distribution

The pore size distribution, the distribution of the porosity corresponding to the pores with different sizes, has a great relationship with the permeability, the consolidation and the strength of clay. The porosity refers to the proportion of the pore areas to the binary image. For the pores with similar diameter, the larger the porosity is, the more percentage of the pores occupies. Thus, it is valuable to study the distribution of the pore size. The increase of cumulative porosity can be divided into three obvious stages: the smooth stage, the slowly increasing stage and the sharply increasing stage, as illustrated in Fig. 5.8. It can be seen that the pore diameter mainly concentrated in the range of 1.0–8.0 μm, that is the sharply increasing stage. Comparing the porosity of the samples before and after freezing-thawing, the porosity of the freezing-thawing clay without loadings is about 11.24% higher than that of the undisturbed clay. Under the same cyclic loadings, the porosity of the freezing-thawing sample with freezing temperature of −10 °C is smaller than those of −20 and −30 °C, though its maximum pore diameter is larger than those of others.

According to the definition of the porosity, the change of porosity indeed reflects the variation of the pore volume within the samples. The distribution of the cumulative porosity versus the axial strain and the excess pore pressure, respectively and are shown in Fig. 5.9. It can be seen that there is no obvious tendency of the porosity distribution with axial strain. Moreover, the porosity distribution with excess pore pressure presents a tendency that the cumulative porosity decreases with the excess pore pressure increasing. Since the axial strain without considering the radial strain cannot take place of the volumetric strain, the development of the excess pore pressure is totally due to the drainage of the pore water which decreases the pore volume of the samples. As a result, the higher the excess pore pressure develops within the saturated soft clay, the smaller the porosity is.

5.3.3.4 Fractal Research on the Microscopic Pore Structures of Silty Clay

There has been a growing interest in fractal analysis of the porous materials including the soft silty clay. Additionally much research work has corroborated that the pore and particle geometry have fractal characteristics at scales ranging from microscopic to macroscopic grains (Tarquis et al. 2006). The obtained fractal dimension D values do not only quantitatively describe the complexity of clay microstructure, but also provide a better understanding of the variation of the microstructure under loading conditions. In this study, the perimeter-area fractal dimension based on the slit island method (Meisel 1991) was employed to describe the variation of the complexity in

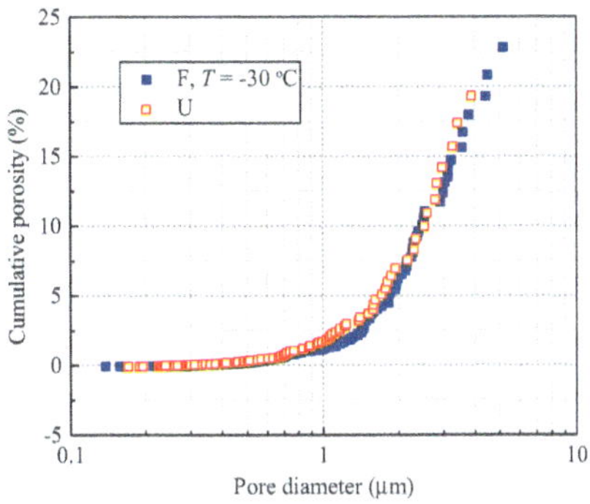

(a) Samples before and after freezing-thawing

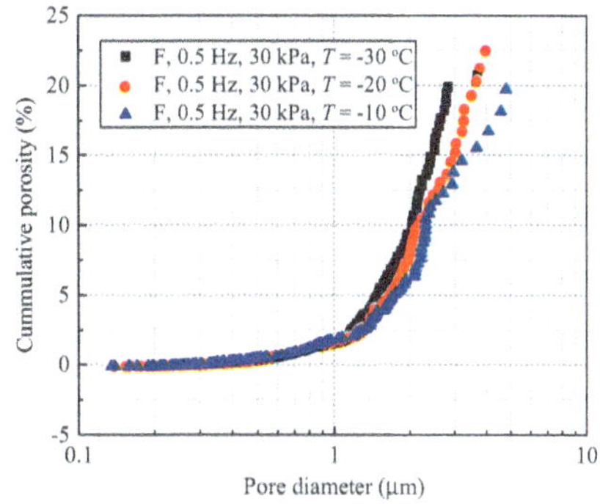

(b) Samples with different freezing temperatures

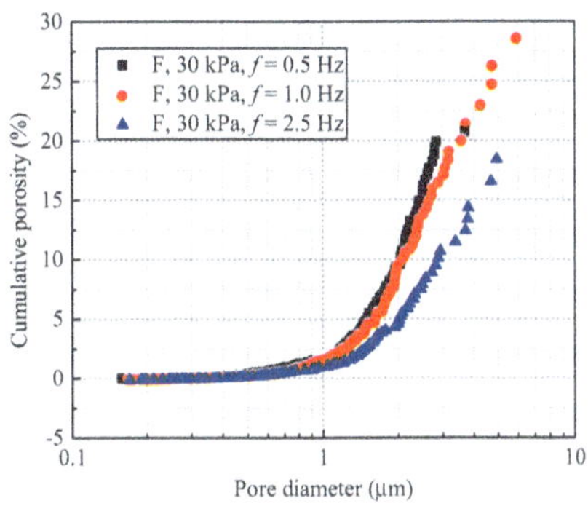

(c) Freezing-thawing samples with different frequencies

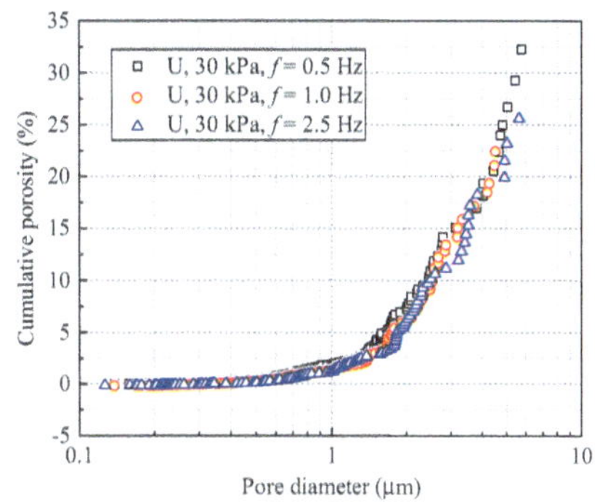

(d) Undisturbed samples with different frequencies

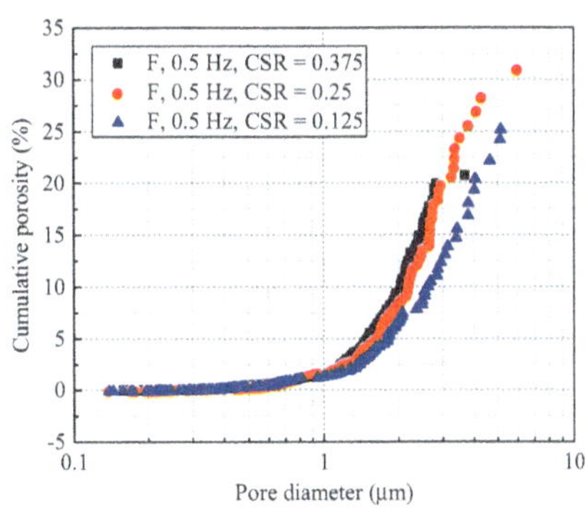

(e) Freezing-thawing samples with different CSRs

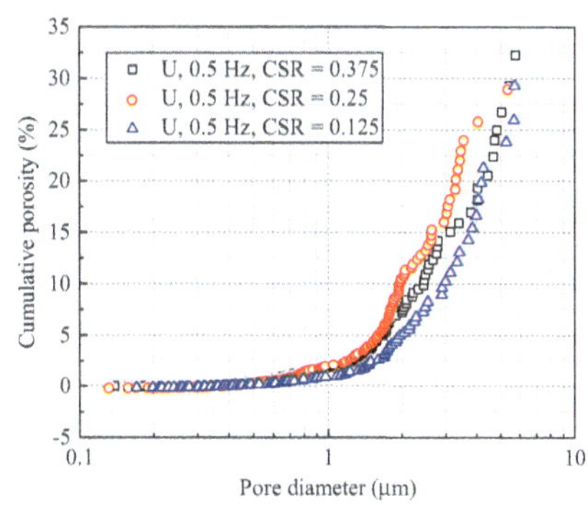

(f) Undisturbed samples with different CSRs

Fig. 5.8 Fitting results of the cumulative curves of porosity

Fig. 5.9 The distribution of the cumulative porosity from SEM images

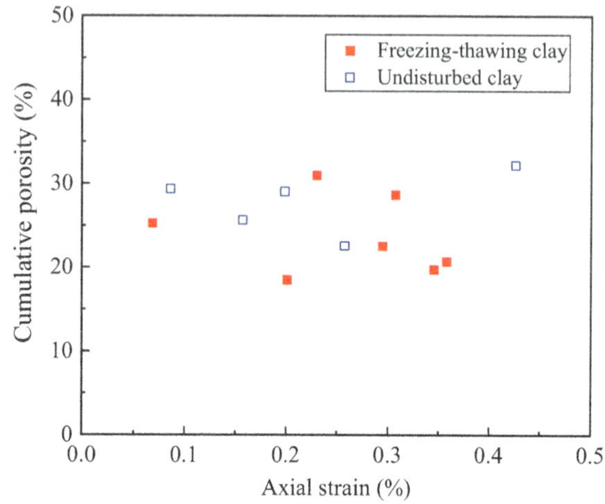

(a) With the axial strain

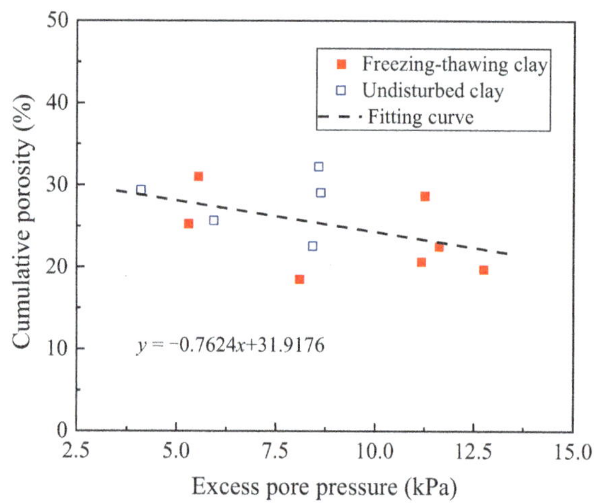

(b) With the excess pore pressure

the microscopic pore structures. The relationship between the perimeter and the area of pores in the SEM images is characterized as

$$\log(P) = D/2 \times \log(A) + C \tag{5.4}$$

where P is the perimeter of the pore; A is the area of the pore; C is the intercept of straight line and D is two times of the line slope in double logarithm coordinate system. Under the cyclic loading, the pores become smaller and part of the pore boundaries are compressed to be closed. The shape of the pores should be more regular, which indicates that the fractal dimension should decrease.

The fractal results of the microscopic pore structure of the silty clay before and after freezing-thawing is illustrated in Fig. 5.10. It can be seen that all samples appear to have fractal characteristics. The fractal dimension values are between [1.15, 1.27], meeting the requirements of fractal dimension within [1.0, 2.0]. The similarity of the microscopic pore structures is reduced by the distortion from the freezing-thawing effects as well as the cyclic loading for the reason that the fractal dimension is related to the roughness of the microstructure of silty clay. The fractal dimension of the freezing-thawing samples decreases with the frequency increase, while the fractal dimension of the undisturbed samples experiences an inflection point at 1.0 Hz. For the samples with different CSRs, the fractal dimension of both freezing-thawing and undisturbed samples increases with the CSR increasing. The lower the frequency and the larger the cyclic stress are, the more energy transferred to samples is. The disturbance from the cyclic loading increases the fractal dimension of the microstructures.

5.4 Discussion

In order to investigate the variation of microstructure of silty clay before and after repeated freezing-thawing cycles, the SEM tests were performed with samples subjected to cyclic loading. Based on the SEM images, the influence of frequency, the cyclic stress ratio and the effect of freezing temperature on the microstructure parameters was analyzed. Additionally, the relationship between the fractal dimension and the macro-parameter was addressed. Meanwhile, appropriate physical limitations are also generated.

There is dispersion among different samples obtained from the field. Thus it is hard to control the same condition when the samples for SEM tests are made. The differences observed between the porosity and the pore size values are related to the dissimilarities among the samples. For the silty clay sample, the presence of two phases has been detected in the SEM test at high magnifications as shown in Fig. 5.1. Most of the pores are irregularly shaped and elongated, and few impurities even remain after sublimation of the water takes place. Because of this, a series of homogeneous samples from the same location is necessary to guarantee statistical significance of the microstructure analysis. The limited size of the sample used for the SEM test, with respect to the large size of the sample used for the triaxial cyclic test, should also be taken into account.

Beside inherent differences among the samples, the test process and the image analysis methods might also result in dispersion. In this study, the effect of magnification from 1000× up to 5000× is not significant in the porosity analysis of SEM images. However, as described before, the SEM test results are strongly influenced by grayscale threshold value. With an appropriate threshold, the soil aggregates and pores can be differentiated truly and reasonably. Otherwise, it will cause large deviation in SEM test results. Since each SEM imagine presents only one section of each

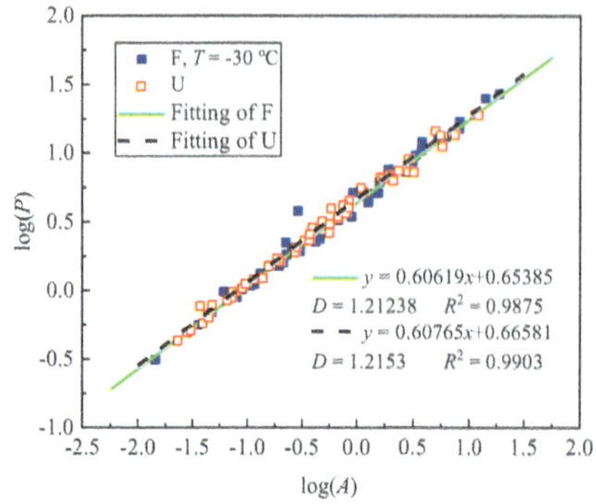

(a) Samples before and after freezing-thawing

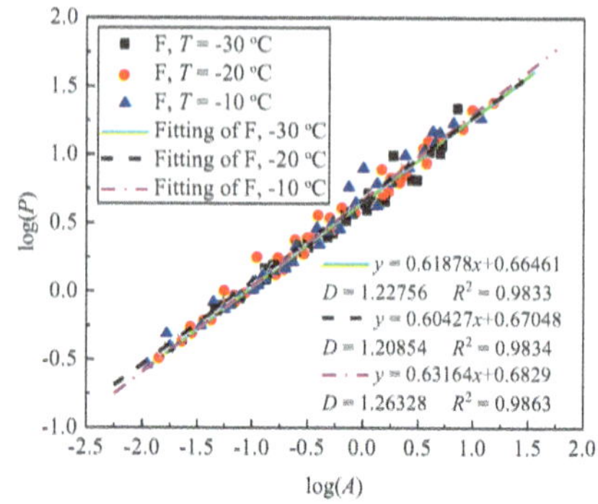

(b) Samples with different freezing temperatures

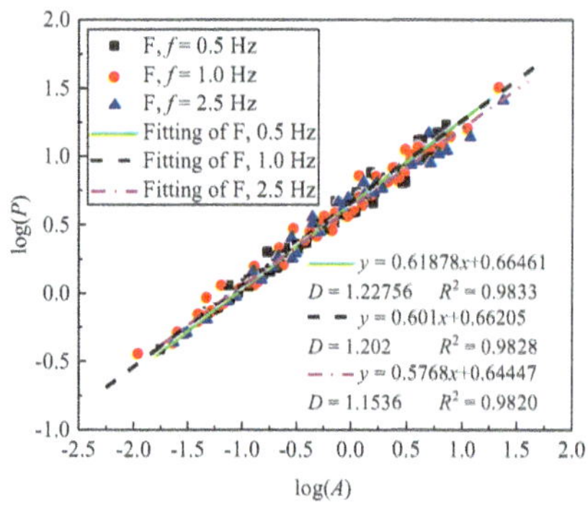

(c) Freezing-thawing samples with different frequencies

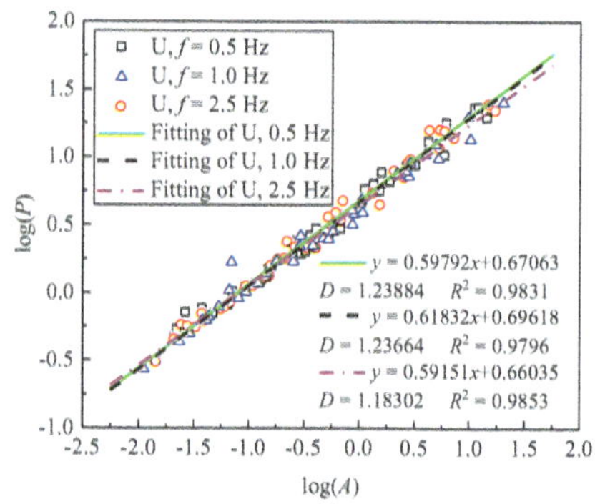

(d) Undisturbed samples with different frequencies

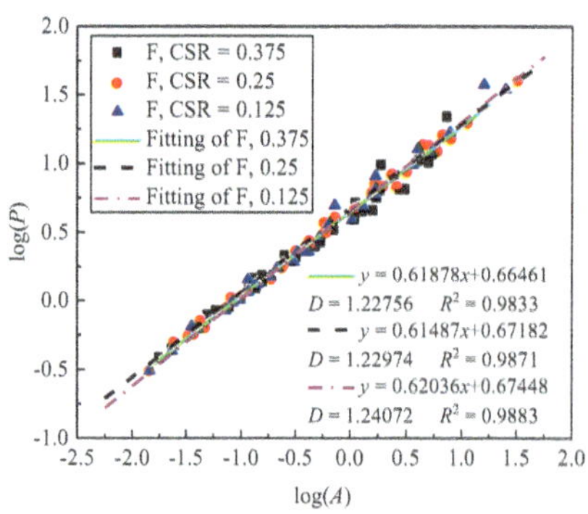

(e) Freezing-thawing samples with different CSRs

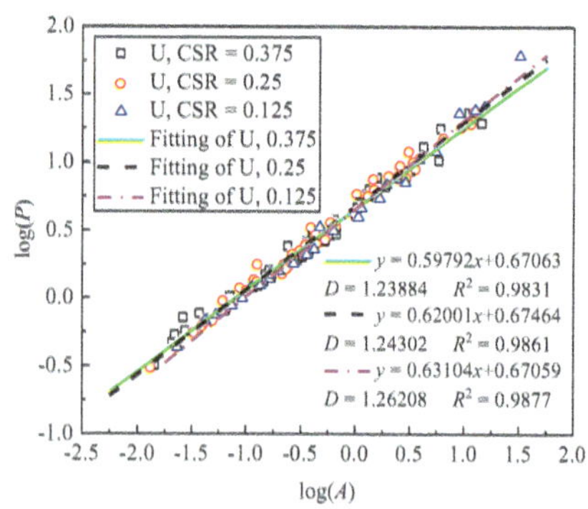

(f) Undisturbed samples with different CSRs

Fig. 5.10 Fitting results of the fractal dimension

sample, the porosity in each section must be equal to the total porosity of each sample. So the curves of soil filling rate versus different thresholds were compared based on the Otsu method. The corresponding threshold to the real filling rate can be back calculated by the polynomial fitting function between the filling rate and threshold. For the silty clay of layer No. 5 in Shanghai, the porosity is estimated by reducing about 30% to the calculated porosity with the Otsu method, and the threshold was selected as [85] for the original SEM images, with error of porosity within 5%.

Figures 5.11a, b illustrate the distribution of the fractal dimension with the axial strain and the excess pore pressure, respectively. The effect induced by cyclic loading has been typically attributed to the development of the axial strain and the excess pore pressure, which represents the energy absorbed by the samples. The fractal dimension

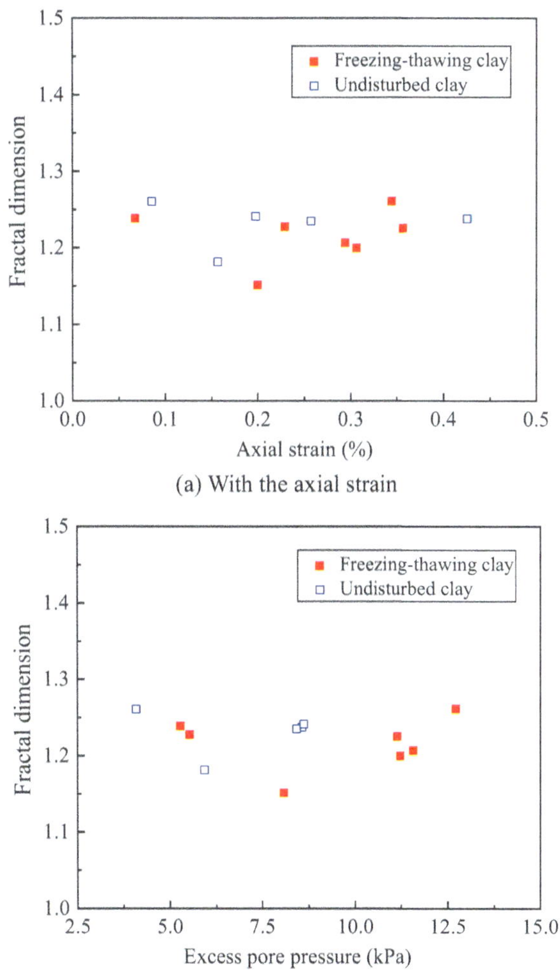

(a) With the axial strain

(b) With the excess pore pressure

Fig. 5.11 The distribution of the fractal dimension from SEM images represents the complexity of the microscopic pore structures. However, there is no obvious relationship between the two behaviors. Figures 5.12a, b present the variation of the fractal dimension with the frequency and the cyclic stress ratio, respectively. It can be seen that the behavior of the freezing-thawing clay was consistent with that of the undisturbed clay. The fractal dimension decreases with the increase in frequency. The fractal dimension of the freezing-thawing samples decreases about 6% from 0.5 to 2.5 Hz. It is for this reason that the cyclic loading with higher frequency transfers less energy to samples, along with smaller fractal dimension. In addition, the fractal dimension of the silty clay decreased with the CSR increasing. The larger the cyclic stress is, the larger the deformation develops (Pillai et al. 2011; Cui and Zhang 2015).

Fig. 5.12 The variation of the fractal dimension of the samples under different cyclic loadings

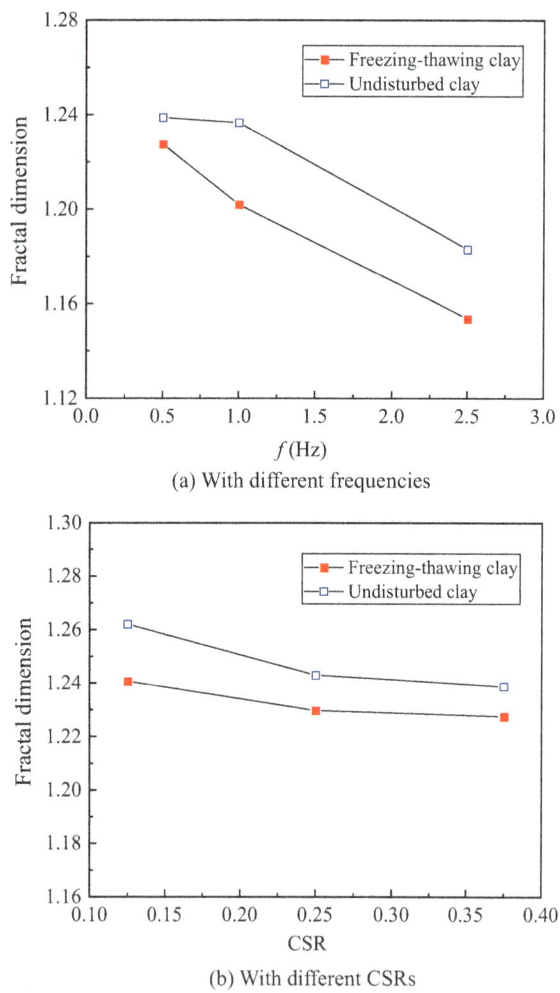

The similarity among the microscopic pore structures with different diameters, as well as the fractal dimension, decreases when subjected to the cyclic loading.

The variation of the microstructure parameters, including the porosity, the pore shape coefficient, the pore anisotropy, and the fractal dimension, was analyzed under different cyclic loadings before and after freezing and thawing. For the SEM tests, each SEM imagine presents only one section of each sample and the analysis is based on each individual pore. Among these parameters, the fractal dimension as a representative statistical parameter, was selected for discussion of the relationships among the microscopic pore structure parameters. For the irregular distribution between the fractal dimension and the axial strain or the excess pore pressure, is mainly resulted from the dispersion of the samples. But with different frequencies or CSR values, the fractal dimension of both freezing-thawing and undisturbed samples shows obvious developing tendencies, which means that the microstructure of the samples was effected by both the freezing-thawing and cyclic loadings. Moreover, the tentative analysis of the relationship between the micro- and macrostructure still need more research.

5.5 Conclusions

The main goal of the current study was to investigate the variation of the microscopic pore structures of the silty clay samples before and after freezing-thawing under the subway vibration loading. The parameters of microscopic pore structure of the silty clay, including the pore shape coefficient, the pore orientation anisotropy, the pore size distribution and the fractal dimension, are evaluated with different frequencies, cyclic stress ratios and freezing temperatures. The main conclusions are obtained as follows.

(1) The corrected Otsu method was used to obtain the threshold for the SEM images of silty clay. The freezing-thawing clay exhibits larger aggregates and larger pore structures than those of the undisturbed clay. After cyclic loading, the microstructures of the particle aggregates are compacted.
(2) The freezing-thawing process has little effect on the pore shape coefficient of the silty clay, but the mean anisotropy value of the microscopic pore structures increases about 12% after freezing-thawing. The pore orientation of samples without loading are arranged horizontally, while the pores of samples under cyclic loadings are arranged vertically.
(3) The porosity increases with the pore diameter increasing in the range of 1.0–8.0 μm. The porosity of samples without loadings increases about 11.24% after freezing-thawing. The lower the freezing temperature is, the larger the porosity of the samples is. There is no obvious tendency of the porosity distribution with the axial strain, while the porosity decreases with the excess pore pressure increasing.

(4) The microscopic pore structures of the silty clay exhibit fractal characteristics with the fractal dimension between [1.15, 1.27]. The fractal dimension is reduced by the disturbance from the freezing-thawing effect, coupled with the cyclic loading.

References

Bird, N., Diaz, M. C., Saa, A., et al. (2006). Fractal and multifractal analysis of pore-scale images of soil. *Journal of Hydrology, 322*(1), 211–219.

Chamberlain, E. J., & Gow, A. J. (1979). Effect of freezing and thawing on the permeability and structure of soils. *Engineering Geology, 13*(1), 73–92.

Cui, Z. D., He, P. P., & Yang, W. H. (2014). Mechanical properties of a silty clay subjected to freezing-thawing. *Cold Regions Science and Technology, 98,* 26–34.

Cui, Z. D., & Jia, Y. J. (2013). Analysis of electron microscope images of soil pore structure for the study of land subsidence in centrifuge model tests of high-rise building groups. *Engineering Geology, 164,* 107–116.

Cui, Z. D., & Zhang, Z. L. (2015). Comparison of dynamic characteristics of the silty clay before and after freezing and thawing under the subway vibration loading. *Cold Regions Science and Technology, 119,* 29–36.

Cui, Z. D., Zhao, L. Z., & Yuan, L. (2016). Microstructures of consolidated Kaolin clay at different depths in centrifuge model tests. *Carbonates and Evaporites, 31*(1), 47–60.

Dathe, A., Eins, S., Niemeyer, J., et al. (2001). The surface fractal dimension of the soil-pore interface as measured by image analysis. *Geoderma, 103*(1), 203–229.

De Boever, W., Derluyn, H., Van Loo, D., et al. (2015). Data-fusion of high resolution X-ray CT, SEM and EDS for 3D and pseudo-3D chemical and structural characterization of sandstone. *Micron, 74,* 15–21.

Ewa, K. M. (2015). SEM investigations of clay subjected to 200 MPa pressure. *Procedia Earth and Planetary Science, 15,* 791–796.

Hattab, M., & Fleureau, J. M. (2010). Experimental study of kaolin particle orientation mechanism. *Geotechnique, 60*(5), 323–331.

Hattab, M., & Fleureau, J. M. (2011). Experimental analysis of kaolinite particle orientation during triaxial path. *International Journal for Numerical and Analytical Methods in Geomechanics, 35*(8), 947–968.

Hemes, S., Desbois, G., Urai, J. L., et al. (2015). Multi-scale characterization of porosity in Boom Clay (HADES-level, Mol, Belgium) using a combination of X-ray μ-CT, 2D BIB-SEM and FIB-SEM tomography. *Microporous and Mesoporous Materials, 208,* 1–20.

Liu, C., Shi, B., Zhou, J., et al. (2011). Quantification and characterization of microporosity by image processing, geometric measurement and statistical methods: Application on SEM images of clay materials. *Applied Clay Science, 54*(1), 97–106.

Liu, J. Y., & Zhang, L. J. (2014). The microstructure characters of saline soil in Qarhan Salt Lake area and its behaviors of mechanics and compressive strength. *Arabian Journal for Science and Engineering, 39*(12), 8649–8658.

Liu, Z., Liang H. C., & Zhang J. (2010). The SEM analysis of rock-soil mini-structure after saturation. In *GeoShanghai 2010 International Conference*. Shanghai, China.

Liu, Z. B., Shi, B., Inyang, H. I., et al. (2005). Magnification effects on the interpretation of SEM images of expansive soils. *Engineering Geology, 78*(1), 89–94.

Meisel, L. V. (1991). Perimeter-area analysis, the slit-island method and the fractal characterization of metallic fracture surfaces. *Journal of Physics. D. Applied Physics, 24*(6), 942.

Ninjgarav, E., Chung, S. G., Jang, W. Y., et al. (2007). Pore size distribution of Pusan clay measured by mercury intrusion porosimetry. *KSCE Journal of Civil Engineering, 11*(3), 133–139.

Othman, M. A., & Benson, C. H. (1993). Effect of freeze-thaw on the hydraulic conductivity and morphology of compacted clay. *Canadian Geotechnical Journal, 30*(2), 236–246.

Otsu, N. (1975). A threshold selection method from gray-level histograms. *IEEE Transactions on Systems, Man, and Cybernetics, 9*(1), 62–66.

Pillai, R. J., Robinson, R. G., & Boominathan, A. (2011). Effect of microfabric on undrained static and cyclic behavior of kaolin clay. *Journal of Geotechnical and Geoenvironmental Engineering, 137*(4), 421–429.

Shi, B., Wu, Z., Inyang, H., et al. (1999). Preparation of soil specimens for SEM analysis using freeze-cut-drying. *Bulletin of Engineering Geology and the Environment, 58*(1), 1–7.

Tang, Y. Q., Zhou, J., Hong, J., et al. (2012). Quantitative analysis of the microstructure of Shanghai muddy clay before and after freezing. *Bulletin of Engineering Geology and the Environment, 71*(2), 309–316.

Tarquis, A. M., McInnes, K. J., Key, J. R., et al. (2006). Multiscaling analysis in a structured clay soil using 2D images. *Journal of Hydrology, 322*(1), 236–246.

Vasseur, G., Djeran, Maigre I., Grunberger, D., et al. (1995). Evolution of structural and physical parameters of clays during experimental compaction. *Marine and Petroleum Geology, 12*(8), 941–954.

Wang, L. B., Frost, J. D., & Lai, J. S. (2004). Three-dimensional digital representation of granular material microstructure from X-ray tomography imaging. *Journal of Computing in Civil Engineering, 18*(1), 28–35.

Wei, X., Hattab, M., Fleureau, J. M., et al. (2013). Micro-macro-experimental study of two clayey materials on drying paths. *Bulletin of Engineering Geology and the Environment, 72*(3), 495–508.

Zhang, L. M., & Li, X. (2010). Microporosity structure of coarse granular soils. *Journal of Geotechnical and Geoenvironmental Engineering, 136*(10), 1425–1436.

Zhang, Z. L., & Cui, Z. D. (2017). Analysis of microscopic pore structures of the silty clay before and after freezing–thawing under the subway vibration loading. *Environmental Earth Sciences, 76*, 528.

Zhou, J., & Tang, Y. Q. (2015). Artificial ground freezing of fully saturated mucky clay: Thawing problem by centrifuge modeling. *Cold Regions Science and Technology, 117*, 1–11.

Chapter 6
Microscopic Pore Structures of Silty Clay Before and After Freezing-Thawing by MIP

6.1 Introduction

With the rapid development of the subway rail transit in Shanghai, the effect of the cyclic vibration loadings on the surrounding foundations and buildings has been an area of intense investigation. The soft clay, with multiscale heterogeneity of structure, is characterized by high water content, high compressibility, high sensitivity, low permeability and low bearing capacity (Cui and Zhang 2015). Generally, the soil mass is mainly constituted of two parts, skeleton particles and pore water. For silty clay, the component of the skeleton particles includes original minerals (feldspar, quartz and mica, etc.), clay minerals (kaolinite, illite, montmorillonite, etc.), soluble salts, and so forth. In particular, the clay minerals play an important role in controlling the characteristics of the soft clay. Flocculated structure and dispersed structure have been found to be two representative structures for clayey soils, which have a profound influence on the mechanical behavior of soil (Prashant and Penumadu 2007; Pillai et al. 2011). However, the complexity of soft clay is determined by not only the mineral composition, but also the pore water inter- or intra-aggregates structures. As the soil becomes finer, the coarse content in the soil descends and the soil structure gradually changes from a coarse controlled structure to a fines-controlled structure which is easily affected by a wetting or drying process (Zhang and Li 2010). With the water content of soil increasing, the soil structure changes from good integration to flocculation that the particles connect to each other mainly by edges and angularities. Irrespective of how complexity of the microstructure was, the reduced total porosity and altered the pore size distribution (PSD) occurred in the clay under the compaction conditions by external dynamic or static loadings (Thom et al. 2007). In addition, the knowledge on the PSD which affects permeability, consolidation, strength and so on is important to the component of the skeleton particles of soft clay, even if both are of considerable interest.

To guarantee the safety in the construction of subway tunnel in soft soil areas, the artificial freezing method has been widely utilized in the construction of by-passes

Z.-D. Cui et al., *Dynamics of Freezing-Thawing Soil around Subway Shield Tunnels*,
https://doi.org/10.1007/978-981-15-4342-5_6

of the subway tunnel due to its unique advantages as both a structural support and a water barrier. In the freezing process, the ice crystal aggregates gradually increased. Chamberlain and Gow (1979) found a reduction in void ratio and an increase in vertical permeability with four fine-grained soils before and after freezing-thawing, which attributed to the formation of polygonal shrinkage cracks. The denser the microstructure is, the stronger the ice crystals aggregate, and soil exhibits higher mechanical and strength characteristics (Liu and Zhang 2014). However, it also brings some problems, such as frost heave, thawing settlement, differential settlement and structural leakage (Tang and Yan 2015). After freezing-thawing, the dynamic elastic modulus descends with the dynamic strain increasing (Cui et al. 2014). Moreover, large thaw settlement appeared in centrifuge model test, with great improvement in permeability after artificial ground freezing-thawing (Zhou and Tang 2015). The changes in these phenomena, including deformations, cracks, excess pore water pressure, and strength degradation, are controlled by the state of the microscope pore structures of soil to a great extent. Therefore, it is of primary importance to conduct accurate measurements of the PSD of the silty clay before and after freezing-thawing.

Recently a growing number of researchers have tried to interpret those phenomena based on the variation of the soil microstructures. The microstructure (<100 μm) refers to the pore structure, crystal structure and mineral composition which can be observed by the scanning electron microscope (SEM), the mercury intrusion porosimetry (MIP), the transmission electron microscopy (TEM), the X-ray diffraction (XRD), the computerized X-ray tomography (CT) and so forth (Ninjgarav et al. 2007; Lubelli et al. 2013; Wei et al. 2013; Zong et al. 2015). Among these technologies, the MIP has been routinely used to quantitatively examine the wide range of pore sizes. The microstructure of soil can be better characterized by a PSD, which can be measured using the MIP (Penumadu and Dean 2000). Initially, the adequacy of the MIP was questioned that whether the intrusion pressure will impose minimal disturbance to the soil fabric or not. In spite of high pressure applied by the MIP, the force acting to cause damage was moderate, particularly in fine pores. In recent years, some researchers have given the data derived from the SEM and the MIP, which indicated that damage from the MIP was negligible (Simms and Yanful 2002, 2004). The results of inter- or intra-aggregates structures by the SEM and the MIP were of the same order of magnitude and the PSD measured by the MIP was close to the actual PSD. MIP has been proposed to be reliable and suitable for study of the soil microstructures.

The MIP tests were carried out by Penumadu and Dean (2000) to evaluate the compression occurred in PSD of cohesive kaolin samples. The tests showed substantive initial compression before the occurrence of actual intrusion resulting in errors associated with the interpretation of pore diameters in the range of 0.4–200 μm. Remarkably differences of the PSD of Pusan clay, with a decline in the inter- and intra-aggregate pore spaces caused by increase in consolidation pressure, were recognized by the mean pore size which effectively represent characteristics of the PSD (Ninjgarav et al. 2007). The same results of compaction effects on the PSD of silty loam were obtained by Lipiec et al. (2012), as well as the effects of aggregate size. It is confirmed from the MIP tests that there exist two major groups of pores within the soft clay, inter- and intra-aggregate pores. It was impossible to find a correspondence between the volume of total porosity and the volume of intra-aggregate pores, though the volume of pore water was found to agree closely with the volume of intra-aggregate pores (Monroy et al. 2010). In the drying process, the inter-aggregate pore water first evaporates, followed by the intra-aggregate pore water. Amongst the inter-aggregate pores, smaller diameter pores are more sensitive to changes in moisture content (Sasanian and Newson 2013). In addition, only the inter-aggregate pores are changed and the intra-aggregate pores remain almost the same during drained triaxial shearing (Yu et al. 2016). The movements, including the rotation and translation of particles and particle aggregates, happened to soil samples under loadings (Cui et al. 2016).

6.2 MIP Test

The MIP tests were carried out using Quantachrome PoreMaster 33 automatic mercury intrusion porosimeter, with two low pressure stations and one high pressure station. It achieves a maximum pressure of 33,000 psi for pore size measurements in the range from over 950 to 0.0064 μm pore diameter. The MIP involves intruding the voids of a porous material with a nonwetting fluid (mercury) under increasing absolute pressures. The intrusion pressure is inversely related to the equivalent pore entry diameter. The theoretical relationship between the intrusion pressure and the pore diameter, assuming a right cylinder with a constant radius, was shown as follows.

$$p = -\frac{2\gamma_m \cos\theta_m}{r} \tag{6.1}$$

where p is intrusion pressure, r being pore radius, γ_m being the surface tension of mercury which has the value of 0.485 N/m at 25 °C and θ_m is the contact angle between mercury and soil interface which has the value of 140° in this study.

By the increasing of the intrusion pressure from p to $p + \delta p$, the corresponding radius of pores filled by mercury will decrease from r to $r - \delta r$. The volume pore size distribution function, differential volume $D_V(r)$, defined as the pore volume per unit interval of pore radius, gives:

$$D_V(r) = \frac{dV}{dr} = -\frac{p}{r}\left(\frac{dV}{dp}\right) \tag{6.2}$$

where V is the normalized volume of intrusion mercury in milliliters per one gram of dried samples. The volume log radius distribution function, log-differential volume $D_V(\log r)$, can be expressed as:

$$D_V(\log r) = \frac{dV}{d\log r} = rD_V(r) \tag{6.3}$$

Combining Eqs. (6.3) and (6.2) gives:

$$D_V(\log r) = -p\frac{dV}{dp} = -\frac{dV}{d\log p} \tag{6.4}$$

Complete dehydration is an essential preparation work for MIP analysis on clay samples. There are several possible methods to remove the water from a clay sample while keeping the fabric of the material invariant, including the air drying, the oven drying, the freeze drying and the critical-point drying. It is generally accepted that "freeze-drying" causes the minimum fabric alterations resulting from the suction effect on clay samples (Ma 2014). The samples were generally frozen rapidly in liquid nitrogen and then freeze-dried in a cold vacuum chamber with condenser temperature about –50 °C for about 24 h. Consequently the ice is sublimated excluding passing through liquid state, avoiding parasitic suction effect (Shi et al. 1999). In this study, the "freeze-drying" method was adopted. In order to ensure that the MIP tests reveal the PSD of the silty clay, preparing and test procedure are as follows.

(1) Take out the sample form the testing apparatus carefully. Cut open the rubber membrane outside with sharp scissors.
(2) Incise a wedge notch along the height of each sample with a sharp blade. Cut the middle part of each sample into strips of approximately 10 × 10 × 20 mm (length × width × height).
(3) Put the strips in liquid nitrogen to be frozen instantaneously. Subsequently put them into the vacuum freeze-drying machine to be thoroughly dried.
(4) Separate the dried strips into certain sizes excluding the fabric disturbed. Some are analytically weighed and placed in the dilatometer, which are then assembled and inserted into the low pressure (0–40 psi) station for intrusion.
(5) After low pressure intrusion, the samples and dilatometer were weighed again, and then inserted into the high pressure (20–33,000 psi) station for intrusion. The rate of pressure increase must be very low and the equilibrium times on each step must be long enough to avoid perturbation of the samples.
(6) Analyze the results after high pressure intrusion and dispose the remaining mercury.

6.3 Test Results and Discussion

6.3.1 Effects of Freezing-Thawing on the PSD

In order to study the effects of the freezing-thawing on the PSD of silty clay, the MIP tests on samples before and after freezing-thawing were carried out (Zhang and Cui 2018). The normalized volumes from the MIP tests against the pore radii of the samples without cyclic loadings before and after freezing-thawing are plotted in

Fig. 6.1. In the MIP tests, the intrusion and extrusion curves show the variations of the mercury volume with the different pore sizes. From Eq. (6.1), it can be obtained that the intrusion pressure is inversely related to the pore radius, corresponding one by one. For the intrusion process, the total mercury intrusion volume increases with the intrusion pressure increasing. It can be seen from Fig. 6.1 that the mercury increases slowly at the beginning of lower pressure, where the mercury mainly fills the pores larger than 1 μm. The cumulative intrusion volume increases rapidly at the pore radii of 0.3–0.6 μm. Moreover, the mercury intrusion becomes a very obvious trend due to substantial pores with size of 0.1–0.5 μm, which indicates that there exists an bottleneck effect in the intrusion stage of the MIP tests (Cui and Tang 2011). When the mercury passes through the bottleneck pores, the cumulative intrusion volume will rise to a larger value with a smaller pressure increment. After that, the pores with size

Fig. 6.1 The intrusion and extrusion curves of samples

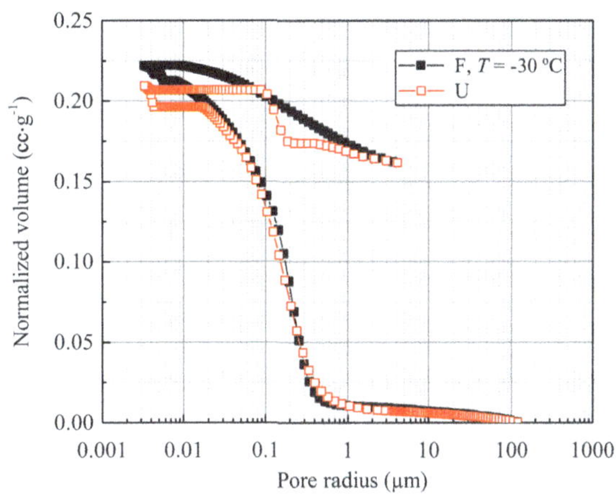

(a) Without cyclic loadings

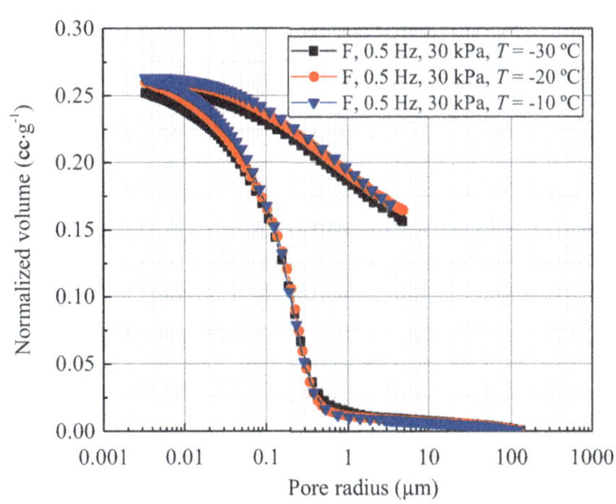

(b) With different freezing temperatures

less than 0.01 μm will be filled by the mercury. However, it is difficult for the mercury to enter the smallest and the closed pores in the samples even when it is nearly the largest pressure. Thus, the mercury intrusion rate is slowly decreasing to the end of intrusion process. For the extrusion process, some of the mercury is trapped in pores with narrow necks which leads to hysteresis of the curves. Actually, the volume of the trapped pores or the remnant mercury stands for the volume of inter-aggregate pores and the total intrusion volume minus the volume of the inter-aggregate pores stands for the volume of intra-aggregate pores.

In Fig. 6.1a, the final mercury intrusion volume of the freezing-thawing samples is about 6.0% larger than that of the undisturbed samples without cyclic loadings, though the intrusion processes are similar to each other. It is for the reason that the water in the saturated soil turns into ice when the temperature is lowered below its freezing point. The volume of soil expands and gradually the pores are full of ice among the soil particles, resulting in the sliding of the particles and the deformation of the pore shapes. After thawing, the ice in the soil changes to be water, which leaves enlargement to the microscope pore structures of the silty clay. In addition, the freezing temperatures also have an effect on the PSD of the silty clay. Figure 6.1b shows variations of the mercury intrusion volume of the samples with different freezing temperatures under the same cyclic loading. The mercury intrusion curves are close to each other before passing through the pores with size larger than 0.1 μm. The final mercury intrusion volume increases with the freezing temperature decreasing. It is considered that the intra-aggregate pores are sensitive to the freezing temperatures, while the inter-aggregate pores are not. The lower the freezing temperature is, the more the film water and free water freeze to ice, exhibiting more mercury intrusion volume. Moreover, the variations of the mercury volume among the samples with different freezing temperatures are extremely small, due to the compaction of cyclic loadings.

Figure 6.2 presents the log-differential mercury volume curves of samples before and after freezing-thawing. The variation of the log-differential mercury volume is the pore size distribution of the samples, where the abscissa values of the peak points are called the most probable pore sizes. It represents that the pores with the same pore sizes appear most within the samples. For the silty clay, the log-differential pore volume curves show a unimodal feature, mainly in the pore radius range from 0.05 to 0.6 μm. However, for the clay with flocculated structures, the differential volume curves may show a bimodal mode (dual-porosity structure), where the mode with larger pore radii represents inter-aggregate pores and the mode with smaller pore radii represents intra-aggregate pores (Zhang and Li 2010). Irrespective of the unimodal or bimodal mode, the mode with smaller pore radii of the curves contain significant portions of the microscope pores. For the samples without loadings, the most probable pore size of undisturbed and freezing-thawing samples are 0.17158 μm and 0.22415 μm, respectively, increasing by 30.64%. For the freezing-thawing samples under the same loading conditions, the most probable pore sizes of samples with different freezing temperatures are close to each other. The reason for that might be that the water in the pores have been fully frozen into ice when the temperature reaches −10 °C. So the effects of freezing temperatures on the pore structures are not

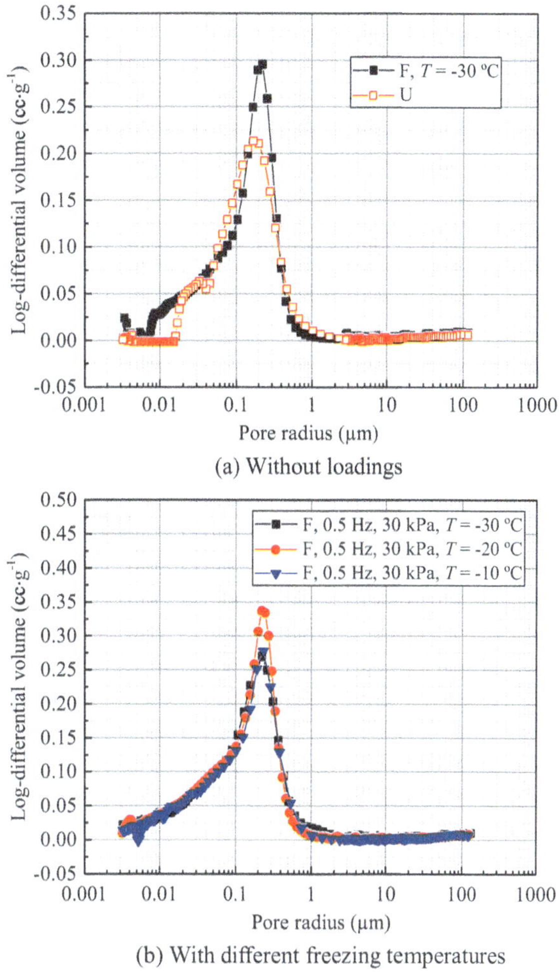

Fig. 6.2 Log-differential pore volume curves of samples

obvious. Besides, the samples are subjected to cyclic stress after thawing. The values of the most probable pore sizes of samples with different freezing temperatures are close to each other.

6.3.2 *Effects of Frequency on the PSD*

Figure 6.3 shows the intrusion and extrusion curves of samples with different frequencies. The volume of mercury gradually increases according to the decrease of the pore radius, along with the increase in mercury intrusion pressure. Similar with

Fig. 6.3 The intrusion and extrusion curves of samples with different frequencies

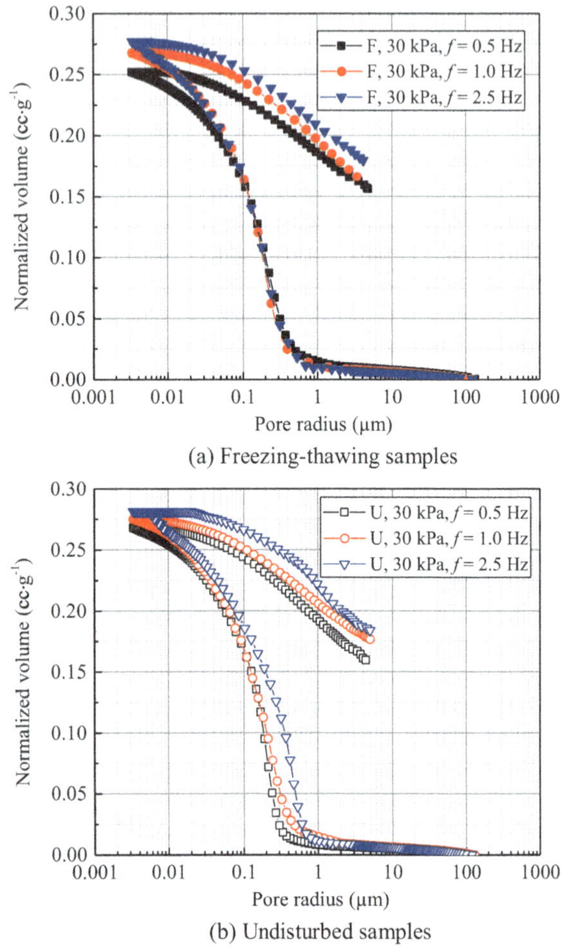

Fig. 6.1, the intrusion curves exhibit bottleneck phenomena, where the mercury volume increases slowly at the beginning and ending of the MIP tests and increases rapidly at the pore radii of 0.2–0.6 μm. When the pore radius is large, the difference among the pore volumes is not very large; whereas, for small pores, it is harder for the mercury to enter, so the difference gradually appears. The trapped mercury in inter-aggregate pores with narrow necks results in misalignments between the intrusion and extrusion curves. The final mercury volume of samples with 2.5 Hz is larger than those with 1.0 and 0.5 Hz. The volume of the pores within the samples increases with the frequency of the cyclic loadings increasing. The lower the frequency of cyclic loadings is, the larger the deformation of the samples is. For freezing-thawing samples, the volume of samples under the cyclic loading with frequency of 2.5 Hz and CSR of 0.375 is 8.74% larger than that with 0.5 Hz, though most of the intrusion curves of freezing-thawing samples with different frequencies are coincident. For

the undisturbed samples, the volume of samples under cyclic loading with frequency of 2.5 Hz and CSR of 0.375 is 5.03% larger than that with 0.5 Hz. Comparing with the undisturbed samples, the mercury volume of the freezing-thawing samples under the same cyclic loadings is larger than that of the undisturbed samples due to the effects of freezing-thawing.

The log-differential mercury volume curves of samples with different frequencies before and after freezing-thawing are presented in Fig. 6.4. It can be seen that there are no great differences among the most probable pore sizes (0.2–0.3 μm) of the freezing-thawing samples under different frequencies. However, the most probable pore sizes of undisturbed samples differ widely under different frequencies. It is hard to analyze the respective effects of freezing-thawing and cyclic loadings on the variation of PSD. The most probable pore size of undisturbed samples with 2.5 Hz is

Fig. 6.4 Log-differential pore volume curves of samples with different frequencies

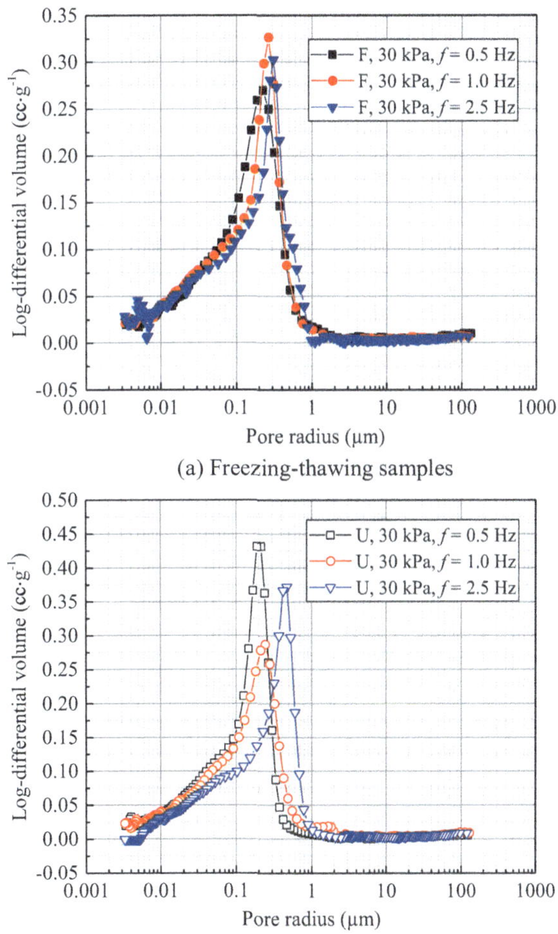

(a) Freezing-thawing samples

(b) Undisturbed samples

about 1.5 times larger than that of 0.5 Hz. The lower the frequency of cyclic loadings is, the more the volume of the pores decreases. The decrease of the most probable pore sizes with the decreasing of the frequency indicates more collapse of pore aggregates within the samples under severe loadings. Similarly, the log-differential pore volume curves of the samples with different frequencies show a unimodal PSD, mainly in the pore radius range from 0.1 to 0.6 μm.

6.3.3 Effects of CSR on the PSD

The evolution of the normalized mercury volume versus pore radius of the silty clay with different CSRs is shown in Fig. 6.5. At the low pressure stage, the mercury mainly fills the larger pores with a slow increase of the mercury intrusion volume. When the mercury overcomes the bottleneck effect, the intrusion volume increases rapidly. However, it is difficult for the mercury to enter the smallest and the closed pores with the size less than 0.01 μm. Thus, the mercury intrusion rate is slowly decreasing to the end of intrusion process. Under the same mercury intrusion pressure (responding to the same pore radii), the final volume of samples with CSR of 0.125 is the maximum. It can be concluded that the sizes of pores are greatly affected by the cyclic loading pressure. Comparing with the undisturbed samples, the pore volumes of the freezing-thawing samples are contracted under the cyclic loadings. However, the mercury intrusion curve of undisturbed samples with the CSR of 0.125 exhibits a sudden increase point with the pore size of 0.007 μm. Moreover, the final mercury intrusion volume is about 27.6% greater than that of undisturbed samples with the CSR of 0.375. The reason for that can be explained by the internal heterogeneous structures of the silty clay. The more homogeneous the PSD of the silty clay is, the smoother the intrusion curve is. The heterogeneous structures in soil samples often lead to the rise and fall of mercury intrusion paths, accompanied by the sudden increase points of mercury pressure.

Figure 6.6 shows the log-differential mercury volume curves of samples with different CSRs before and after freezing-thawing, where the intra-aggregate pores with the radius of 0.1–0.6 μm bring a unimodal PSD to the silty clay. For the freezing-thawing clay, the most probably pore size of the samples with CSR of 0.125 is about 81.2% larger than that with CSR of 0.375. The result is similar to the undisturbed clay. It confirms the general conclusions that pore sizes are related to the loading pressures. As the cyclic stress increases, the deformation of the samples increases and the cumulative intrusion volume increment becomes smaller. Thus, the radii of the most frequent pores tend to descend.

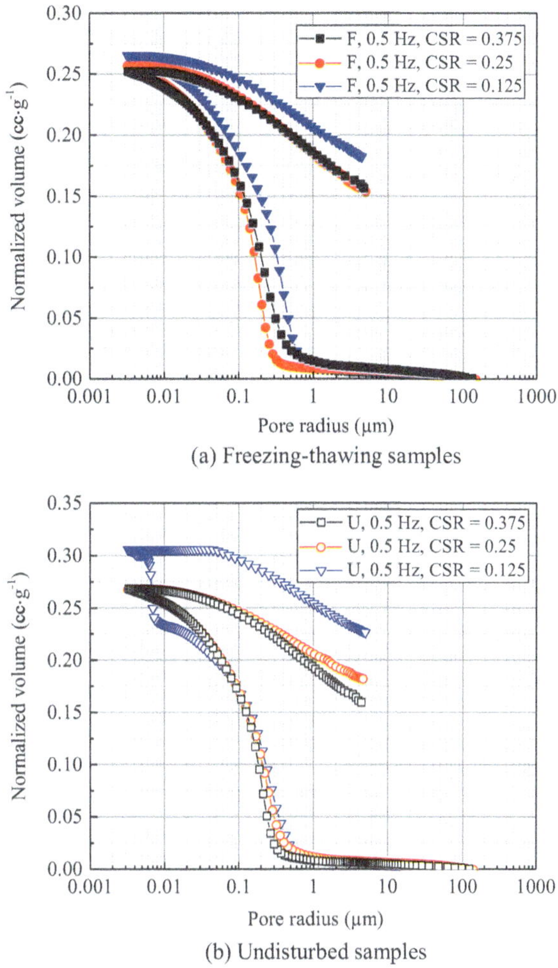

Fig. 6.5 The intrusion and extrusion curves of samples with different CSRs

6.3.4 *Prediction of Water Retention Curve*

Water retention curve (WRC), also called the soil-water characteristic curve (SWCC), is defined as the relationship between the water content and the suction. The characterization of water retention behavior is fundamental to a comprehensive understanding and modelling of the coupled water flow and deformation in unsaturated soils (Hu et al. 2013). It is used to predict the soil water storage and soil aggregate stability. Since water is stored in the soil pores and flows through inter-connected pore chains, soil hydraulic properties are controlled by the soil microstructures (Fredlund et al. 1994). Attempt has been made to relate the soil microstructures to the soil water characteristics. For instance, similar WRC models incorporate the microstructure parameters of the soil were proposed or improved to describe the water retention

Fig. 6.6 Log-differential pore volume curves of samples with different CSRs

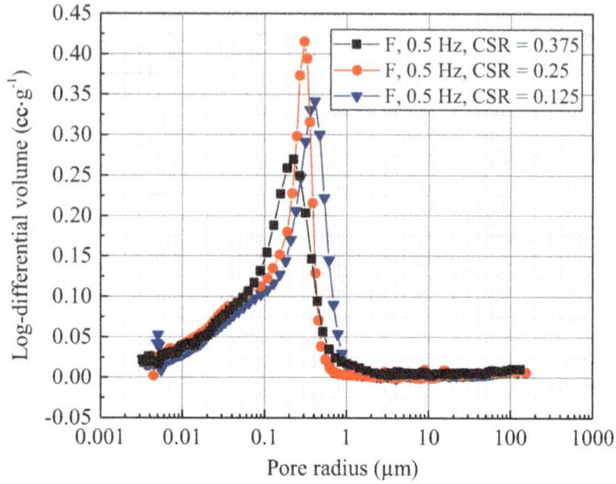

(a) Freezing-thawing samples

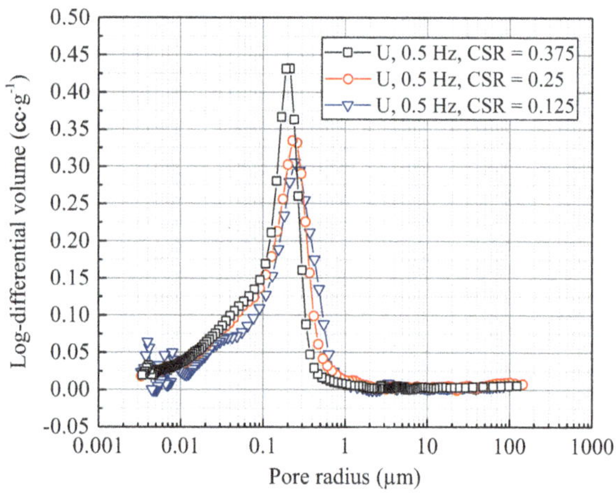

(b) Undisturbed samples

behaviors of soil (Van Genuchten 1980; Durner 1994; Fredlund and Xing 1994). Neglecting the effect of deformation on the water content in laboratory tests, the models are capable of reproducing the dependence of water content on suction with good accuracy for soils, even if the WRC is characteristic for different types of soil (Zhai and Rahardjo 2015; Otalvaro et al. 2016).

For unsaturated soil, the most popular and commonly used water retention model is the van Genuchten model that the PSD index parameters are introduced to represent the influence of the PSD on soil water retention behaviors. The shape of water retention curves can be characterized by:

$$w(s) = w_r + \frac{w_s - w_r}{\left[1 + (\alpha|s|)^n\right]^{1-1/n}} \tag{6.5}$$

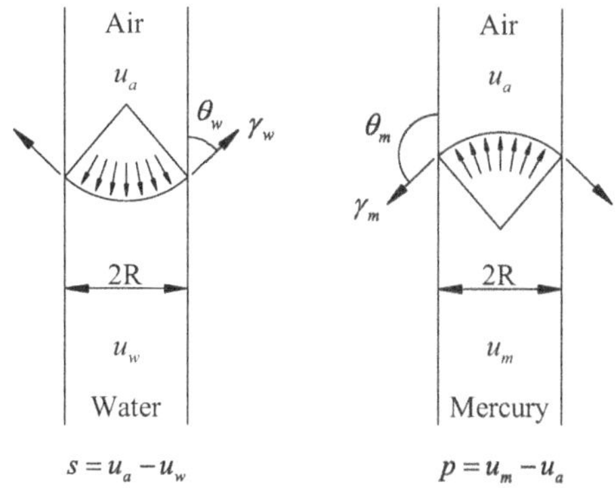

Fig. 6.7 Air injected in drying WRC and mercury intrusion in the cylindrical pore

where w is the water content, w_r being the residual water content, w_s being the saturated water content, s being the suction pressure, α being related to the inverse of the air entry suction, and n being a measure of the PSD.

Since the clay samples were thoroughly dehydration before the MIP tests, assuming the residual water content to be zero, the fitting model was written as follows.

$$w(s) = \frac{w_s}{\left[1 + (\alpha|s|)^n\right]^{1-1/n}} \tag{6.6}$$

Mercury intrusion in a porous medium is a process similar to air injection in a saturated soil as described along the drying water retention curve (Aung et al. 2001). Therefore, the PSD of soil plays a decisive role in its water retention behavior. Figure 6.7 presents an analogy between the processes of air injection in drying WRC and mercury intrusion in the cylindrical pore. It is generally recognized that the soil suction can be related to the intrusion pressure, as well as the water content to the pore volume. Based on the PSD derived from the MIP tests, the curves between the water content and the suction were predicted with the van Genuchten model.

For the unsaturated soil, the matrix suction is equal to the air pressure minus the water pressure.

$$s = u_a - u_w \tag{6.7}$$

where u_a is the air pressure and u_w is the water pressure. The intrusion pressure is equal to the mercury pressure minus the air pressure.

$$p = u_m - u_a \tag{6.8}$$

where u_m is the mercury pressure.

Similar to Eq. (6.1), the suction can be obtained in Fig. 6.7 as:

$$s = \frac{2\gamma_w \cos\theta_w}{r} \tag{6.9}$$

where γ_w is the surface tension of water which has the value of 0.072 N/m at 25 °C and θ_w is contact angle between water and soil interface which is assumed to be 0° in this study.

If the pore radii in Eqs. (6.1) and (6.9) are the same, the two pressures are related by the following:

$$s = 0.194p \tag{6.10}$$

It should be stressed that the mercury intrusion process is similar to the drying process of soil. Therefore, the intrusion volume of mercury is equivalent to the air volume within the soil as the follows.

$$V_m = V_a \tag{6.11}$$

where V_m is the intrusion mercury volume and V_a is the air volume.

In MIP tests, the normalized volume can be expressed as follows.

$$V = \frac{V_m}{m_s} \tag{6.12}$$

where m_s is the weight of the dried clay samples for the MIP.

In drying process of saturated clay, the water content is written as

$$w = \frac{m_w}{m_s} = \frac{(V_V - V_a)\rho_w}{m_s} = w_s - \frac{V_a\rho_w}{m_s} \tag{6.13}$$

where m_w is the weight of the water in the soils, V_V being the total volume of pores in clay, and ρ_w is the density of water which is approximately 1.0 g/cm^3.

Combining Eqs. (6.11)–(6.13) gives:

$$w = w_s - \frac{V_m\rho_w}{m_s} = w_s - V \tag{6.14}$$

The MIP derived water content and the measured water content of the samples were compared by Sasanian and Newson (2013). Below 50%, the MIP derived and the measured water contents are equal, while higher than 50%, the MIP derived water content are lower than the expected one. For the silty clay used in this study, the natured water content is below 50% and the MIP derived water content can be reliable. With the dry bulk density and total porosity obtained from the MIP tests, the saturated content of the clay sample can be calculated as follows.

$$w_s = \frac{S_r n \rho_w}{\rho_d} = \frac{S_r n \rho_w}{(1-n)\rho_s} \tag{6.15}$$

where S_r is the degree of saturation which is assumed to be 1, n being the total porosity, ρ_d being the dry bulk density, and ρ_s is the apparent density in Table 6.1.

Altogether, Eqs. (6.10) and (6.14) allow for the conversion of the mercury intrusion pressure and the normalized volume into the soil suction and the water content, respectively. The van Genuchten model expressed as Eq. (6.6) was adopted for prediction of the water retention curves of the silty clay. It can be seen from Figs. 6.8, 6.9 and 6.10 that the MIP derived water content versus the matrix suction can be well fitted by the van Genuchten model. The prediction results are summarized in Table 6.2.

Table 6.1 Summary of MIP test results

No.	Soils	Dry bulk density (g/cc)	Apparent density (g/cc)	Total porosity (%)	Total intruded volume (g/cc)	Total surface area (m^2/g)	Throat /pore ratio	Most probable pore size (μm)
01	F (−30 °C)	1.608	2.6107	40.5517	0.2522	10.5944	0.0475	0.22444
02	F (−30 °C)	1.524	2.5120	40.8675	0.2681	11.7544	0.1257	0.26467
03	F (−30 °C)	1.595	2.6146	40.5941	0.2775	11.5910	0.0723	0.30771
04	F (−30 °C)	1.326	1.9747	34.2073	0.2580	11.9174	0.0698	0.30011
05	F (−30 °C)	1.394	2.1707	37.1252	0.2663	9.6029	0.1380	0.40671
06	F (−20 °C)	1.588	2.6219	41.3404	0.2603	11.1012	0.0808	0.22616
07	F (−10 °C)	1.559	2.4136	37.1714	0.2634	9.6377	0.1704	0.23534
08	F (−30 °C)	1.428	2.0445	31.7191	0.2221	10.3686	0.7996	0.22415
09	U	1.437	2.2956	38.5307	0.2682	11.6714	0.0552	0.18997
10	U	1.500	2.3618	38.1418	0.2753	10.9854	0.2063	0.24251
11	U	1.538	2.4430	38.7088	0.2817	6.5252	0.4111	0.47114
12	U	1.260	1.8679	33.8586	0.2687	11.1743	0.1433	0.22643
13	U	1.501	2.7090	46.0381	0.3067	27.5275	0.0059	0.24251
14	U	1.456	2.0544	30.5094	0.2095	10.7257	0.0297	0.17158

Fig. 6.8 The water retention curves of samples with different freezing temperatures

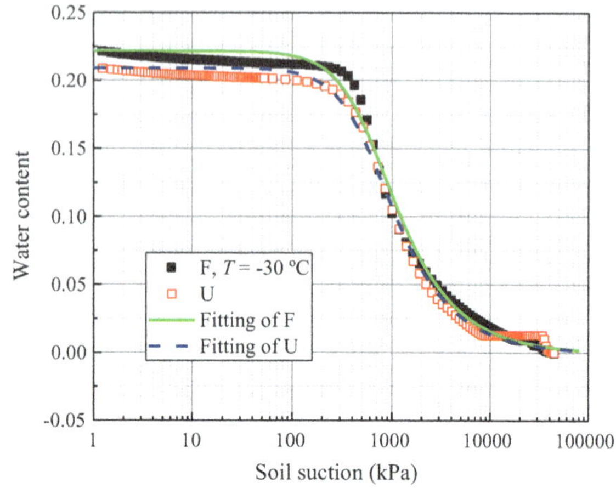

(a) Without cyclic loading

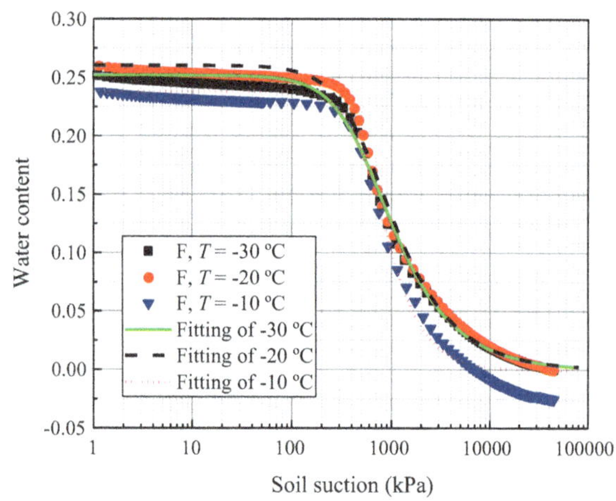

(b) With different freezing temperatures

Similar to the bottleneck effects in the mercury intrusion curves, the water content primarily decreases slowly according to the increase of suction. When reaching a certain value whose range is about 400–600 kPa, it begins to decrease rapidly. At the end stage, the water content decline rate begins to fall, which appears obviously in the van Genuchten model curves. From Figs. 6.1, 6.3 and 6.5, it is observed that the pore radius range of the bottlenecks is about 0.02–0.6 μm, where the mercury intrusion volume increases rapidly. The most frequent pores, corresponding to the most probably pore sizes, provide the most pore volume within the samples. There is a general consensus that the suction of the soil is due to the porosity and the nature of the bonding in the soil. When the suction is close to zero, the soil is close to saturation and the water is held by the soil suction primarily. If the obtained cumulative curves become flat at extremely small suction, the total MIP measured pore volume will

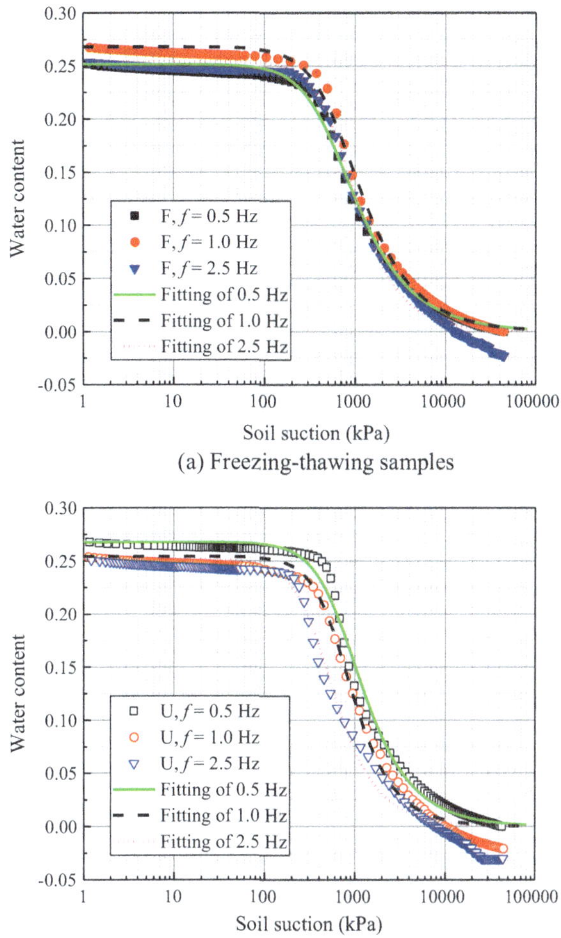

Fig. 6.9 The water retention curves of samples with different frequencies

correspond to the volume of all of the pores within the samples. As water content decreases, binding of the water becomes stronger. The water is strongly bound in the smallest of pores, at contact points among particles, as film water bound around particles by adsorptive forces. The results of the MIP derived WRC illustrate that the MIP tests are highly suited for the prediction of the soil-water characteristic curves.

6.3.5 Analysis of the PSD Using Fractal Theory

Recently the fractal characterization of porous materials, including the soft silty clay, as an applicable and potential tool has been well documented. Extensive research

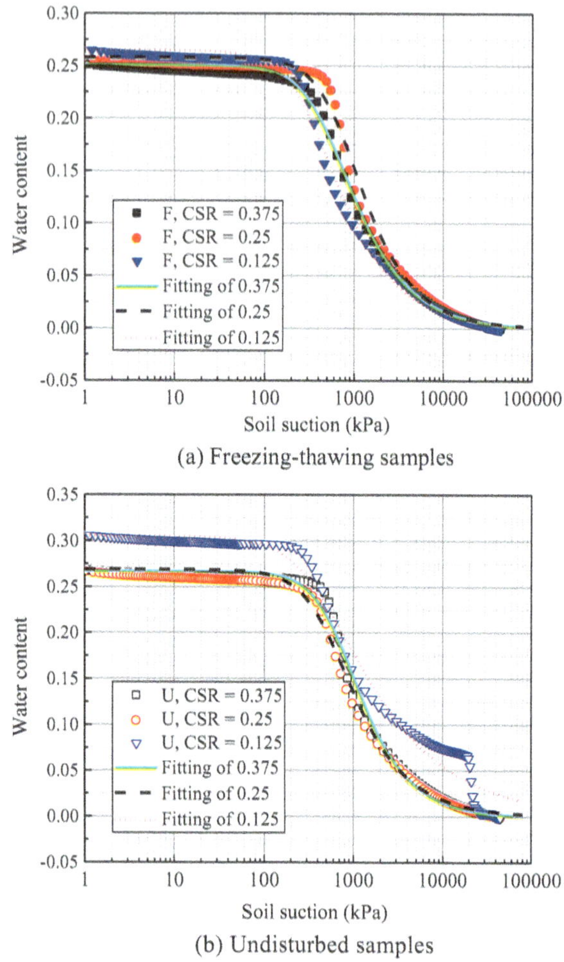

Fig. 6.10 The water retention fitting curves of samples with different CSRs

work has corroborated that both the pores and particles have geometry fractal characteristics at scales ranging from microscopic to macroscopic grains (Tarquis et al. 2006). The obtained fractal dimension D values are not only a quantitative descriptor of the complexity of clay microstructure, but also can provide a better understanding of the variation of the microstructure under loading conditions (Tripathi et al. 2012). Both the soil aggregates and the pore structures can be analyzed by the fractal. For microscope pore structures, the most popular used method for the PSD is the MIP tests. Zhang and Li (1995) proposed the thermal fractal model to obtain the fractal dimensions of porous media from mercury intrusion data. Thereafter this model was chosen for the capillary condensation tests by Wang and Li (1997), comparing the results with other fractal models. In this study, the thermal fractal dimension model

Table 6.2 Summary of the water retention fitting curves in MIP tests

No.	Soils	f (Hz)	CSR	Saturated water content (%)	a	n	R2
1	F (−30 °C)	0.5	0.375	25.22	6.20007E-06	1.9095	0.99898
2	F (−30 °C)	1	0.375	26.82	2.74423E-06	1.9785	0.99915
3	F (−30 °C)	2.5	0.375	25.45	3.72803E-07	2.2719	0.99672
4	F (−30 °C)	0.5	0.25	25.80	1.20924E-06	2.0550	0.99722
5	F (−30 °C)	0.5	0.125	26.63	1.71642E-05	1.8648	0.99785
6	F (−20 °C)	0.5	0.375	26.03	4.62236E-06	1.9395	0.99816
7	F (−10 °C)	0.5	0.375	23.84	2.88539E-08	2.6734	0.99606
8	F (−30 °C)	–	–	22.21	4.65508E-06	1.9299	0.99791
9	U	0.5	0.375	26.81	1.22317E-06	2.0706	0.99734
10	U	1	0.375	25.43	2.43832E-07	2.3524	0.99685
11	U	2.5	0.375	25.17	1.11704E-06	2.2942	0.99259
12	U	0.5	0.25	26.88	4.55648E-06	1.9442	0.99852
13	U	0.5	0.125	30.67	8.46283E-05	1.5405	0.98559
14	U	–	–	20.95	4.05613E-06	1.9548	0.99843

was adopted, following the recommendations of Cui et al. (2016), to analyze the fractal characteristics of the silty clay from the MIP tests.

In the MIP tests, the work to the mercury is equal to the surface energy increment of the mercury liquid. According to the transform of the energy in the cylinder pores, the relationship between the intrusion pressure and the intrusion volume turns out to be:

$$\int_0^V PdV = -\int_0^S \gamma \cos\theta dS \tag{6.16}$$

where S is the surface area of pores. Equation (6.16) is equivalent to:

$$\sum_{i=1}^{n} \overline{P_i}\Delta V_i = kr_n^2(V_n^{1/3}/r_n)^D \tag{6.17}$$

where $\overline{P_i}$ is the mean intrusion pressure for ith scattered intrusion; ΔV_i is the intrusion volume for ith scattered intrusion; n is the total number of scattered intrusion minus 1; r_n is the pore radius for nth scattered intrusion; k is constant; V_n is the cumulative intrusion volume for nth scattered intrusion.

If it is given that:

$$W_n = \sum_{i=1}^{n} \overline{P_i} \Delta V_i \tag{6.18}$$

$$Q_n = V_n^{1/3} / r_n \tag{6.19}$$

where W_n and Q_n are the surface energy and the incremental intrusion volume of mercury in MIP tests, respectively.

Substitution of Eqs. (6.18) and (6.19) into Eq. (6.17) yields:

$$\lg\left(W_n / r_n^2\right) = D \lg Q_n + C \tag{6.20}$$

where C is constant. The slope of double logarithmic curve of Q_n and $\left(W_n / r_n^2\right)$, D, is the thermal fractal dimension of samples.

Figure 6.11 shows the thermal fractal results of the MIP tests of the silty clay before and after freezing-thawing. It turns out that the thermal fractal dimension model is appropriate to all samples. The values of thermal fractal dimension are in the reasonable range of 2–3. In addition, the values of the thermal fractal fitting correlation coefficient are greater than 0.98 which means values of thermal fractal dimension are reliable. The similarity of the microscope pore structures is reduced by the distortion involving the freezing-thawing effect, the cyclic loading and even the combination of the both. However, the relationship between thermal fractal dimensions of samples and corresponding cyclic frequencies, CSRs and freezing temperatures is not quite consistent. The fractal dimension of samples without cyclic loadings decreases after freezing-thawing. In contrast to the freezing-thawing samples, the fractal dimension of the undisturbed samples increases with the increase of the frequency. For the samples with different CSRs, the fractal dimension of both freezing-thawing and undisturbed samples decreases with the CSR increasing. Thus, much more researches should be conducted to fully understand the existent relationships between the micro- and macro parameters, though the fractal dimension is related to the roughness of the PSD of the soils.

6.4 Conclusions

In this chapter, 14 MIP tests following the cyclic triaxial tests of the silty clay of layer No. 5 in Shanghai were conducted to investigate the variation of the PSD of the silty clay. The effects of freezing-thawing and cyclic loadings on the PSD of the silty clay were evaluated and the MIP derived water retention curves of the silty clay were predicted with the van Genuchten model. The main conclusions are obtained as follows.

(1) The mercury intrusion pressure is inversely related to the pore radius of the soil. There exists a bottleneck effect in the mercury intrusion process. The mercury volume increases slowly at both the beginning and the ending of the MIP tests

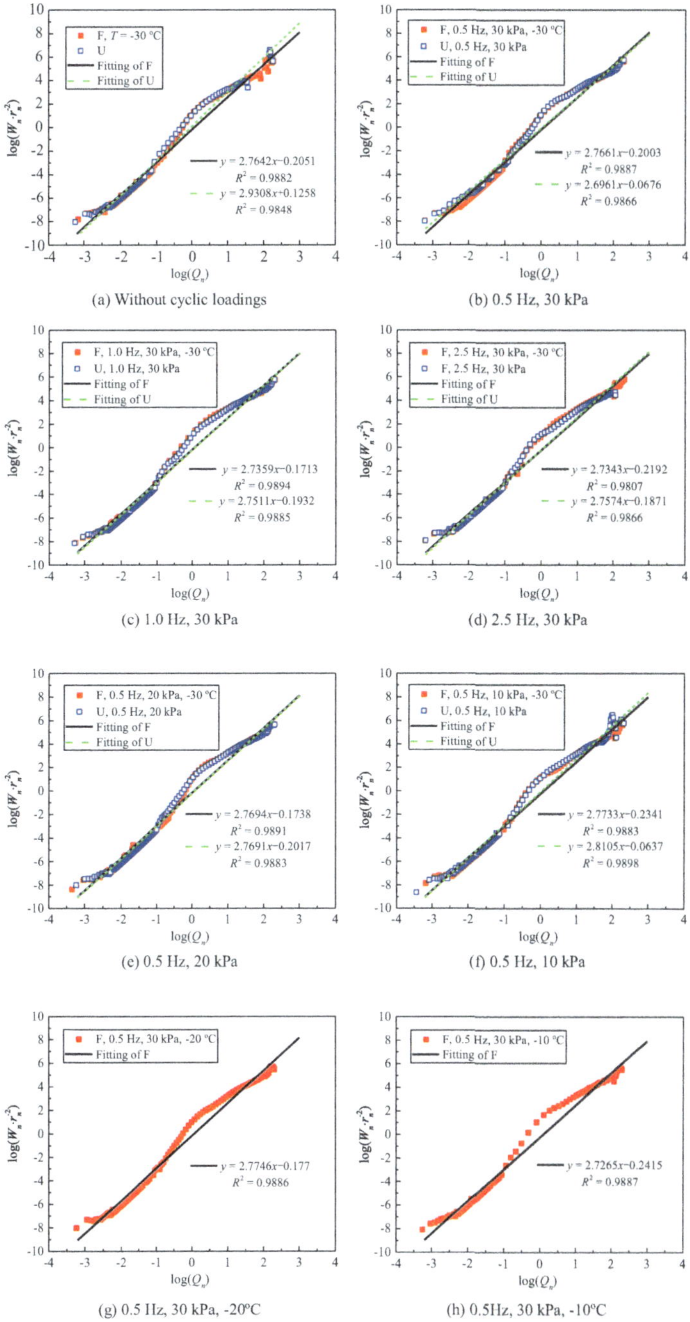

Fig. 6.11 Results of thermodynamics fractal

and increases rapidly at the pore radii of 0.1–0.6 μm. In the extrusion process, some of the mercury is trapped in pores with narrow necks which leads to hysteresis of the curves. The log-differential pore volume curves of the silty clay show a unimodal mode, where the mode with pore radii range from 0.05 to 0.6 μm represents intra-aggregate pores.

(2) The water in the saturated soil turns into ice with the volume expanding when the temperature is lowered below its freezing point. After thawing, the ice in the soil changes to be water, which leaves enlargement to the microscope pore structures of the silty clay. For the samples without cyclic loadings, the final mercury intrusion volume and the most probable pore size increase about 6.0% and 30.64% after freezing-thawing, respectively. Under the same cyclic loading, the mercury intrusion volume increases with the freezing temperature decreasing and the most probable pore sizes are close to each other.

(3) The cyclic loadings have great effects on the PSD of the silty clay. Under the cyclic loading pressure, the pore volumes are contracted. The lower the frequency and the larger the CSR of cyclic loadings are, the larger the deformations of the samples are and the smaller the most probable pore size becomes. The pore volume of freezing-thawing samples with the frequency of 2.5 Hz is 8.74% larger than that with 0.5 Hz, while it is 5.03% of undisturbed samples. The most probably pore size of the freezing-thawing samples with the CSR of 0.125 is about 81.2% larger than that of the CSR of 0.375.

(4) The process of mercury intrusion in a porous medium is similar to air injection in a saturated soil as described along the drying water retention curve. The mercury intrusion pressure and the normalized volume are equivalent to the soil suction and the air volume within the soil, respectively. The MIP derived WRC are suitable for the prediction of the soil-water characteristic curves with the van Genuchten model. In addition, the microscope pore structures of the silty clay exhibit fractal characteristics with the thermal fractal dimension model in MIP tests.

References

Aung, K. K., Rahardjo, H., Leong, E. C., et al. (2001). Relationship between porosimetry measurement and soil-water characteristic curve for an unsaturated residual soil. *Geotechnical and Geological Engineering, 19,* 401–416.

Chamberlain, E. J., & Gow, A. J. (1979). Effect of freezing and thawing on the permeability and structure of soils. *Engineering Geology, 13*(1), 73–92.

Cui, Z. D., He, P. P., & Yang, W. H. (2014). Mechanical properties of a silty clay subjected to freezing-thawing. *Cold Regions Science and Technology, 98,* 26–34.

Cui, Z. D., & Tang, Y. (2011). Microstructures of different soil layers caused by the high-rise building group in Shanghai. *Environmental Earth Sciences, 63*(1), 109–119.

Cui, Z. D., & Zhang, Z. L. (2015). Comparison of dynamic characteristics of the silty clay before and after freezing and thawing under the subway vibration loading. *Cold Regions Science and Technology, 119,* 29–36.

Cui, Z. D., Zhao, L. Z., & Yuan, L. (2016). Microstructures of consolidated Kaolin clay at different depths in centrifuge model tests. *Carbonates and Evaporites, 31*(1), 47–60.
Durner, W. (1994). Hydraulic conductivity estimation for soils with heterogeneous pore structure. *Water Resources Research, 30*(2), 211–223.
Fredlund, D. G., & Xing, A. Q. (1994). Equations for the soil-water characteristic curve. *Canadian Geotechnical Journal, 31*(4), 521–532.
Fredlund, D. G., Xing, A. Q., & Huang, S. Y. (1994). Predicting the permeability function for unsaturated soils using the soil-water characteristic curve. *Canadian Geotechnical Journal, 31*(4), 533–546.
Hu, R., Chen, Y. F., Liu, H. H., et al. (2013). A water retention curve and unsaturated hydraulic conductivity model for deformable soils: Consideration of the change in pore-size distribution. *Geotechnique, 63*(16), 1389–1405.
Lipiec, J., Hajnos, M., & Swieboda, R. (2012). Estimating effects of compaction on pore size distribution of soil aggregates by mercury porosimeter. *Geoderma, 179–180,* 20–27.
Liu, J. Y., & Zhang, L. J. (2014). The microstructure characters of saline soil in Qarhan Salt Lake area and its behaviors of mechanics and compressive strength. *Arabian Journal for Science and Engineering, 39*(12), 8649–8658.
Lubelli, B., de Winter, D. A. M., Post, J. A., et al. (2013). Cryo-FIB-SEM and MIP study of porosity and pore size distribution of bentonite and kaolin at different moisture contents. *Applied Clay Science, 80–81,* 358–365.
Ma, H. (2014). Mercury intrusion porosimetry in concrete technology: tips in measurement, pore structure parameter acquisition and application. *Journal of Porous Materials, 21*(2), 207–215.
Monroy, R., Zdravkovic, L., & Ridley, A. (2010). Evolution of microstructure in compacted London Clay during wetting and loading. *Geotechnique, 60*(2), 105–119.
Ninjgarav, E., Chung, S. G., Jang, W. Y., et al. (2007). Pore size distribution of Pusan clay measured by mercury intrusion porosimetry. *KSCE Journal of Civil Engineering, 11*(3), 133–139.
Otalvaro, I. F., Neto, M. P. C., Delage, P., et al. (2016). Relationship between soil structure and water retention properties in a residual compacted soil. *Engineering Geology, 205,* 73–80.
Penumadu, D., & Dean, J. (2000). Compressibility effect in evaluating the pore-size distribution of kaolin clay using mercury intrusion porosimetry. *Canadian Geotechnical Journal, 37*(2), 393–405.
Pillai, R. J., Robinson, R. G., & Boominathan, A. (2011). Effect of microfabric on undrained static and cyclic behavior of kaolin clay. *Journal of Geotechnical and Geoenvironmental Engineering, 137*(4), 421–429.
Prashant, A., & Penumadu, D. (2007). Effect of microfabric on mechanical behavior of kaolin clay using cubical true triaxial testing. *Journal of Geotechnical and Geoenvironmental Engineering, 133*(4), 433–444.
Sasanian, S., & Newson, T. A. (2013). Use of mercury intrusion porosimetry for microstructural investigation of reconstituted clays at high water contents. *Engineering Geology, 158,* 15–22.
Shi, B., Wu, Z., Inyang, H., et al. (1999). Preparation of soil specimens for SEM analysis using freeze-cut-drying. *Bulletin of Engineering Geology and the Environment, 58*(1), 1–7.
Simms, P. H., & Yanful, E. K. (2002). Predicting soil-water characteristic curves of compacted plastic soils from measured pore-size distributions. *Geotechnique, 52*(4), 269–278.
Simms, P. H., & Yanful, E. K. (2004). A discussion of the application of mercury intrusion porosimetry for the investigation of soils, including an evaluation of its use to estimate volume change in compacted clayey soils. *Geotechnique, 54*(6), 421–426.
Tang, Y. Q., & Yan, J. J. (2015). Effect of freeze-thaw on hydraulic conductivity and microstructure of soft soil in Shanghai area. *Environmental Earth Sciences, 73*(11), 7679–7690.
Tarquis, A. M., McInnes, K. J., Key, J. R., et al. (2006). Multiscaling analysis in a structured clay soil using 2D images. *Journal of Hydrology, 322*(1), 236–246.
Thom, R., Sivakumar, R., Sivakumar, V., et al. (2007). Pore size distribution of unsaturated compacted kaolin: the initial states and final states following saturation. *Geotechnique, 57*(5), 469–474.

Tripathi, S. K., Kushwaha, C. P., & Basu, S. K. (2012). Application of fractal theory in assessing soil aggregates in Indian tropical ecosystems. *Journal of Forestry Research, 23*(3), 355–364.
Van Genuchten, M. T. (1980). A closed-form equation for predicting the hydraulic conductivity of unsaturated soils. *Soil Science Society of America Journal, 44*(5), 892–898.
Wang, F. M., & Li, S. F. (1997). Determination of the surface fractal dimension for porous media by capillary condensation. *Industrial and Engineering Chemistry Research, 36*(5), 1598–1602.
Wei, X., Hattab, M., Fleureau, J. M., et al. (2013). Micro-macro-experimental study of two clayey materials on drying paths. *Bulletin of Engineering Geology and the Environment, 72*(3), 495–508.
Yu, C. Y., Chow, J. K., & Wang, Y. H. (2016). Pore-size changes and responses of kaolinite with different structures subject to consolidation and shearing. *Engineering Geology, 202,* 122–131.
Zhai, Q., & Rahardjo, H. (2015). Estimation of permeability function from the soil-water characteristic curve. *Engineering Geology, 199,* 148–156.
Zhang, B. Q., & Li, S. F. (1995). Determination of the surface fractal dimension for porous media by mercury porosimetry. *Industrial and Engineering Chemistry Research, 34*(4), 1383–1386.
Zhang, L. M., & Li, X. (2010). Microporosity structure of coarse granular soils. *Journal of Geotechnical and Geoenvironmental Engineering, 136*(10), 1425–1436.
Zhang, Z. L., & Cui, Z. D. (2018). Effects of freezing-thawing and cyclic loading on pore size distribution of silty clay by mercury intrusion porosimetry. *Cold Regions Science and Technology, 145*, 185–196.
Zhou, J., & Tang, Y. Q. (2015). Artificial ground freezing of fully saturated mucky clay: Thawing problem by centrifuge modeling. *Cold Regions Science and Technology, 117,* 1–11.
Zong, Y. T., Yu, X. L., Zhu, M. X., et al. (2015). Characterizing soil pore structure using nitrogen adsorption, mercury intrusion porosimetry, and synchrotron-radiation-based X-ray computed microtomography techniques. *Journal of Soils and Sediments, 15*(2), 302–312.

Chapter 7
Dynamics of Silty Clay Around Subway Shield Tunnel Before and After Freezing and Thawing by Numerical Simulation

7.1 Introduction

In recent years, with the rapid increase of population and traffic jam, subway construction has already become an important way to develop underground space and ease traffic jam. In the process of subway construction, especially in some soft soil areas, artificial freezing method has become a common temporary reinforcement method of the foundation. For example, 95% of the connecting passages in Shanghai were constructed by artificial freezing method. Moreover, artificial freezing method is frequently used in the repair for water inflow and collapse in shield tunnel. While after the construction of subway tunnel by artificial freezing method, there is a certain amount of residues of frozen soil, e.g. the remanent frozen soil is about 3–4 m in thickness around the crossing-river tunnel of East Fuxing Road in Shanghai.

The dynamic response of ordinary soil around the tunnel under vibration load has been studied by many researchers by means of various methods, for example microscopic approach (Tang et al. 2014; Yan et al. 2013). While in general, current researches are mainly concentrated on theoretical analysis and numerical simulation. For relatively simple analytical methods, the whole system is mainly simplified as that different kinds of loads move on the Winkler or the half-space foundation (Dieterman and Metrikine 1997; Kaynia et al. 2000). A more complicated and accurate analytical methodology (PiP model) that takes into account the major dynamic characteristics of a three-dimension system for calculating subway vibration was developed by Forrest and Hunt (2006). Moreover, the PiP model was coupled to the track via different supports (Hussein and Hunt 2007). Sheng et al. (2002) presented a method that depends on the boundary integral equations of only the displacement Green's function based on the discrete wavenumber fictitious force method. Lu et al. (2007) presented another model for calculating seismic dynamic response. The model accounted for the problem with piecewise liners embedded in a porous medium. The soil surrounding the tunnel was supposed to be saturated porous medium and was described by Biot's

Z.-D. Cui et al., *Dynamics of Freezing-Thawing Soil around Subway Shield Tunnels*,
https://doi.org/10.1007/978-981-15-4342-5_7

poroelastic theory, and the tunnel was supposed to be the combination of liner pieces and connecting joints which were treated as curved beams and characterized by curved beam theories. Guan and Moore (1994) solved the 3D wave equations for two cylindrical cavities in an unbounded medium to account for the interaction between two deep side-by-side circular tunnels.

For numerical simulation, Israil and Ahmad (1989), Ewing et al. (1958) and Wolf (1985, 1988) concentrated on the dynamic response of soil under the point or line loads. Many researchers also studied the influences of various factors. Lieb and Sudret (1998), Alabi (1992), Yang et al. (2003) and Eason (1965) studied the effect of train speed and depth of soil layer. Rieckh et al. (2012) discussed the dynamic characteristics of layered anisotropic soil under vibration loads. Nejati et al. (2012) studied the influence of subway vibration on the surface through numerical methods and theoretical analysis. Vinson et al. (1978 and 1980) considered the dynamic performance of thawing clay and thawing sandy soil. Ting et al. (1983) researched the dynamic strength of frozen sandy soil based on a large number of statistical data. Vibration from underground railways is frequently modeled by the coupled Finite Element–Boundary Element (FE–BE) technique. Sheng et al. (2006) incorporated the discrete wavenumber method to improve the computational efficiency. Clouteau et al. (2005) developed a numerical model based on FE-BE model for calculating vibration response of subway.

After the construction of artificial freezing, the thawing soil was formed after the melting of the frozen soil and the physical and mechanical properties of soil changed a lot during freezing and thawing. The freezing and thawing weakens the connecting forces between soil particles (Konrad 1989) and increases the permeability of soil (Viklander and Eigenbrod 2000). It also affects the strength and the compressibility of soil (Brams and Yao 1964; Graham and Au 1985). After the freezing and thawing, the permeability of clay (Yang and Zhang 2002) is approximately 3–10 times than that of the undisturbed soil, while the unconfined compressive strength is just about 33–55% of that of undisturbed soil. The deterioration of soil characteristics will lead to the differences between the dynamic response of thawing soil and undisturbed soil. The thawing soil may be more susceptible to be damaged under the same vibration load, so the kinetic properties and deformation of thawing soil around the subway tunnel are of vital significance for the local stability of the tunnel. But the researches on the dynamic characteristics of thawing soil are little, only a small amount focuses on the dynamic response under seismic load (Singh and Donovan 1977; Finn and Yong 1978).

In this chapter, the three-dimension governing equations were derived and the solving method was given. The material parameters of the thawing soil and the undisturbed soil used in the numerical simulations were obtained by dynamic triaxial tests. The dynamic response of thawing soil and undisturbed soil in the same environmental conditions were compared and analyzed (He and Cui 2015).

7.2 Governing Equations

7.2.1 Formulations of the Model

In order to simplify the establishment of model equations, the model in this chapter is based on the following hypotheses: (1) the strain of the soil and tunnel is small enough so that elastic relation can be applied; (2) the properties of the model remain unchanged in the longitudinal direction, that is the model can be treated as a plane strain problem; (3) the interface of the tunnel and the soil satisfies the deformation compatibility.

The governing equations of the whole model consist of three parts: the tunnel, the soil and their interface. It is assumed that the long-term displacement fields of the model are known. The dynamic perturbation displacements $\mathbf{u}$ due to dynamic loads will emerge, which are sufficiently small so that linear equations can be allowed.

The dynamic strain $\boldsymbol{\varepsilon}$ can be expressed by the perturbation displacements $\mathbf{u}$ as follows:

$$\boldsymbol{\varepsilon} = \frac{1}{2}[\nabla \mathbf{u} + (\nabla \mathbf{u})^{\mathrm{T}}] \tag{7.1}$$

According to Hooke's law, the dynamic stress $\boldsymbol{\sigma}$ can be calculated by

$$\boldsymbol{\sigma} = \lambda(\nabla \cdot \mathbf{u})E_3 + 2\mu\boldsymbol{\varepsilon} \tag{7.2}$$

where λ and μ are the Lame constants and they are expressed as $\lambda = \frac{\upsilon E}{(1+\upsilon)(1-2\upsilon)}$ and $\mu = \frac{E}{2(1+\upsilon)}$, in which υ and μ are Poisson's ratio and shear modulus, respectively. E_3 is the 3×3 unit matrix.

The governing equations and boundary conditions of the system in time-space domain can be written as follows.

For the soil:

$$\rho_s \frac{\partial^2 \mathbf{u}_\mathbf{s}}{\partial t^2} = (\lambda + \mu)\nabla(\nabla \cdot \mathbf{u}_\mathbf{s}) + \mu\nabla^2\mathbf{u}_\mathbf{s} + \mathbf{f}_\mathbf{s} \tag{7.3}$$

For the tunnel:

$$\rho_t \frac{\partial^2 \mathbf{u}_\mathbf{t}}{\partial t^2} = (\lambda + \mu)\nabla(\nabla \cdot \mathbf{u}_\mathbf{t}) + \mu\nabla^2\mathbf{u}_\mathbf{t} + \mathbf{f}_\mathbf{t} \tag{7.4}$$

For their interface:

$$\mathbf{u}_\mathbf{s} = \mathbf{u}_\mathbf{t} \tag{7.5}$$

$$\boldsymbol{\sigma}_\mathbf{s} = \boldsymbol{\sigma}_\mathbf{t} \tag{7.6}$$

7.2.2 Equation Solution

As the forms of Eqs. (7.3) and (7.4) are the same, only Eq. (7.3) is solved following. The component forms of Eq. (7.3) in x, y and z direction can be written as:

$$\begin{aligned}\rho_s\frac{\partial^2 u_{sx}}{\partial t^2} &= (\lambda+\mu)\left(\frac{\partial^2 u_{sx}}{\partial x^2}+\frac{\partial^2 u_{sy}}{\partial x\partial y}+\frac{\partial^2 u_{sz}}{\partial x\partial z}\right)+\mu\left(\frac{\partial^2 u_{sx}}{\partial x^2}+\frac{\partial^2 u_{sx}}{\partial y^2}+\frac{\partial^2 u_{sx}}{\partial z^2}\right)+f_{sx}\\ \rho_s\frac{\partial^2 u_{sy}}{\partial t^2} &= (\lambda+\mu)\left(\frac{\partial^2 u_{sx}}{\partial x\partial y}+\frac{\partial^2 u_{sy}}{\partial y^2}+\frac{\partial^2 u_{sz}}{\partial y\partial z}\right)+\mu\left(\frac{\partial^2 u_{sy}}{\partial x^2}+\frac{\partial^2 u_{sy}}{\partial y^2}+\frac{\partial^2 u_{sy}}{\partial z^2}\right)+f_{sy}\\ \rho_s\frac{\partial^2 u_{sz}}{\partial t^2} &= (\lambda+\mu)\left(\frac{\partial^2 u_{sx}}{\partial x\partial z}+\frac{\partial^2 u_{sy}}{\partial y\partial z}+\frac{\partial^2 u_{sz}}{\partial z^2}\right)+\mu\left(\frac{\partial^2 u_{sz}}{\partial x^2}+\frac{\partial^2 u_{sz}}{\partial y^2}+\frac{\partial^2 u_{sz}}{\partial z^2}\right)+f_{sz}\end{aligned} \tag{7.7}$$

The component forms of Eqs. (7.5) and (7.6) can be expressed as:

$$u_{sx}=u_{tx}\quad u_{sy}=u_{ty}\quad u_{sz}=u_{tz} \tag{7.8}$$

$$\begin{aligned}\sigma_{sx}=\sigma_{tx}\quad \sigma_{sy}=\sigma_{ty}\quad \sigma_{sz}=\sigma_{tz}\\ \tau_{sxy}=\tau_{txy}\quad \tau_{syz}=\tau_{tyz}\quad \tau_{szx}=\tau_{tzx}\end{aligned} \tag{7.9}$$

Equation (7.7) is transformed by Laplace transformation for the time t and Fourier transformation for x, y and z, that is:

$$\tilde{u}(s,\omega_x,\omega_y,\omega_z)=\int_0^{+\infty}\int_{-\infty}^{+\infty}\int_{-\infty}^{+\infty}u(t,x,y,z)e^{-st}e^{-i\omega_x x}e^{-i\omega_y y}e^{-i\omega_z z}\mathrm{d}t\mathrm{d}x\mathrm{d}y\mathrm{d}z$$

The Eq. (7.7) can be transformed as:

$$\begin{aligned}\rho_s s^2\tilde{u}_x &= -(\lambda+\mu)\omega_x(\omega_x\tilde{u}_x+\omega_y\tilde{u}_y+\omega_z\tilde{u}_z)-\mu(\omega_x^2+\omega_y^2+\omega_z^2)\tilde{u}_x+\tilde{f}_{sx}\\ \rho_s s^2\tilde{u}_y &= -(\lambda+\mu)\omega_y(\omega_x\tilde{u}_x+\omega_y\tilde{u}_y+\omega_z\tilde{u}_z)-\mu(\omega_x^2+\omega_y^2+\omega_z^2)\tilde{u}_y+\tilde{f}_{sy}\\ \rho_s s^2\tilde{u}_z &= -(\lambda+\mu)\omega_z(\omega_x\tilde{u}_x+\omega_y\tilde{u}_y+\omega_z\tilde{u}_z)-\mu(\omega_x^2+\omega_y^2+\omega_z^2)\tilde{u}_z+\tilde{f}_{sz}\end{aligned} \tag{7.10}$$

Three vectors are defined as: $U=\left(\tilde{u}_x,\tilde{u}_y,\tilde{u}_z\right)^{\mathrm{T}}$, $\Omega=\left(\omega_x,\omega_y,\omega_z\right)^{\mathrm{T}}$, $F=\left(\tilde{f}_{sx},\tilde{f}_{sy},\tilde{f}_{sz}\right)^{\mathrm{T}}$. Equation (7.10) can be simplified as:

$$\rho s^2 U=-(\lambda+\mu)\Omega\Omega^{\mathrm{T}}U-\mu\Omega^{\mathrm{T}}\Omega U+F \tag{7.11}$$

Then U can be calculated as follows:

$$U=[\rho s^2 E_3+(\lambda+\mu)\Omega\Omega^{\mathrm{T}}+\mu\Omega^{\mathrm{T}}\Omega]^{-1}F \tag{7.12}$$

where E_3 is the 3×3 unit matrix.

Therefore, the displacement fields in time-space domain can be obtained by inverse Fourier transformation and inverse Laplace transformation with the contact conditions considered. Then the strain fields and stress fields can be obtained through Eqs. (7.1) and (7.2).

7.3 Numerical Model

7.3.1 Model Size

The numerical model in this chapter is based on Subway Line 8 in Shanghai, 8–18 m in depth from the center of the tunnel to the ground surface. The external and inner diameters are 6.2 m and 5.5 m, respectively. The characteristics of the tunnel were assumed to be the same, therefore a plane strain problem was considered and only one interception was chosen, as shown in Fig. 7.1. In order to compare the dynamic response of thawing soil and undisturbed soil, the numerical model was applied to simulate them, respectively. As the aim is to investigate the general differences of dynamic response between thawing soil and ordinary soil, a single kind of soil (thawing soil or undisturbed soil) was assumed around the tunnel to simplify the model. A, B and C in the model were taken as the reference points, as shown in Fig. 7.1.

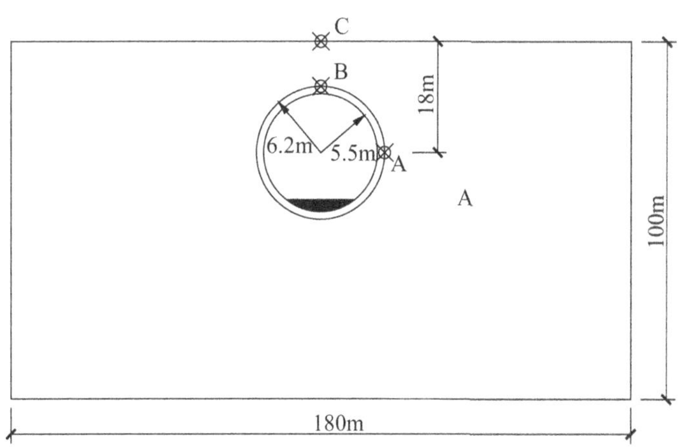

Fig. 7.1 Model size and the reference points

Fig. 7.2 The curve of the vibration force

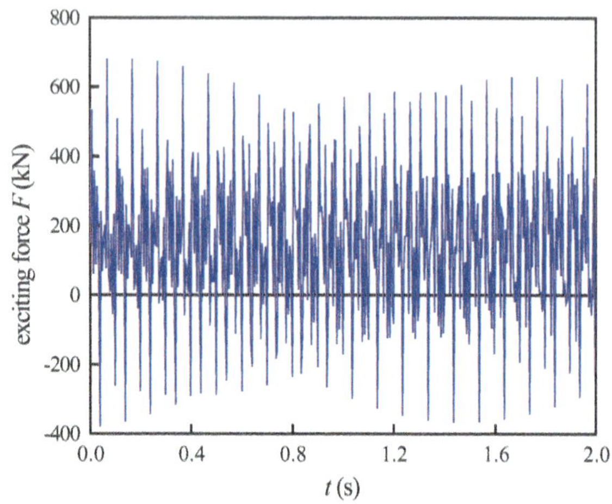

7.3.2 Vibration Load of the Train

According to a lot of researches and tests of the Center of British Railway (Jenkins et al. 1974; Dawn and Stanworth 1979), the vertical force of the train vibration occurs mainly in the three ranges of frequency: the low frequency (0.5–5 Hz), the medium frequency (30–60 Hz) and the high frequency (200–400 Hz). The dominating frequency is the low frequency. According to the frequency ranges of the load, the function of an exciting force can be adopted to represent the vibration load (Pan and Pande 1984):

$$F(t) = A_0 + A_1 \sin \omega_1 t + A_2 \sin \omega_2 t + A_3 \sin \omega_3 t \tag{7.13}$$

where A_1 is the static load of the wheel, A_2, A_3, A_4 are the peaks of the vibration loads which are corresponding to the circular frequencies ω_1, ω_2, ω_3, respectively.

According to the frequency ranges of the load, $A_0 = 1.5 \times 10^5$ N, $A_1 = 1.5 \times 10^5$ N, $A_2 = 1.6 \times 10^5$ N and $A_3 = 2.4 \times 10^5$ N. The corresponding circular frequencies are $\omega_1 = 126$ rad/s, $\omega_2 = 314$ rad/s and $\omega_3 = 503$ rad/s. The vibration time was 2 s. Based on Eq. (7.13), the curve of the vibration force is shown in Fig. 7.2.

7.3.3 Parameters of Soil

The undisturbed soil was dark green silty clay. The undisturbed soil was frozen to −20 °C, and then the thawing soil was obtained after the melting of the frozen soil in a pool with room temperature. The dynamic modulus of the two kinds of soil samples were both obtained by dynamic triaxial tests.

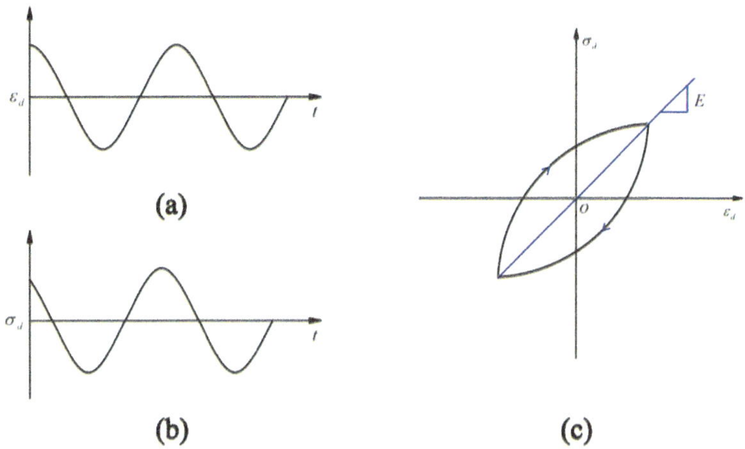

Fig. 7.3 The dynamic stress (**a**), the dynamic strain (**b**) and the hysteresis curve (**c**)

Table 7.1 Parameters of soil

Parameters	Soil		
	Undisturbed soil	Thawing soil	Segment
E (MPa)	270	250	35000
μ	0.35	0.46	0.25
ρ (kg/m^3)	1870	2100	2500

Because the soil is not an ideal elastomer, the dynamic stress (σ_d) and its corresponding dynamic strain (ε_d) are not synchronized, as shown in Fig. 7.3a, b. If the dynamic stress and the dynamic strain in a cycle are painted in the same coordinate plane, the hysteresis curve as shown in Fig. 7.3c can be gained. The average slope of the hysteresis curve is taken as the dynamic modulus E_d:

$$E_d = \sigma_{d\max}/\varepsilon_{d\max} \tag{7.14}$$

In the vibration of underground railway, the maximum strain is less than 10^{-4}, thus the dynamic modulus is assumed to be constant. The soil parameters obtained by the tests are summarized in Table 7.1.

7.3.4 Damping Parameter and Boundary Conditions

Since the presence of the damping and the radiation attenuation caused by the increment of propagation distance, waves will not propagate unlimited in soil medium. Damping is an important attribute of dielectric material, which cannot be ignored. In order to quantify the damping, Rayleigh damping was adopted.

Rayleigh damping can be expressed as:

$$[C] = \alpha[M] + \beta[K] \tag{7.15}$$

where $[M]$ is the mass matrix; $[K]$ is the stiffness matrix; α and β are the damping coefficients.

If the damping ratio ξ_i remains unchanged in a certain frequency range $\omega_i \sim \omega_j$, according to the orthogonal condition of vibration mode, the damping coefficients can be determined:

$$\begin{aligned} \alpha &= \frac{2\omega_i\omega_j}{\omega_i + \omega_j}\xi \\ \beta &= \frac{2}{\omega_i + \omega_j}\xi \end{aligned} \tag{7.16}$$

In order to determine the damping coefficients, two circular frequencies ω_i and ω_j are needed. The general idea is to carry out the analysis of the vibration mode and two modes are selected to calculate the damping coefficients. Because the vibration of the subway is mainly focused on the vertical vibration, two modes of the maximum participated mass of the vertical vibration were selected to determine the damping coefficients.

Above all, the analysis of the vibration mode was carried on. The front 200 modes of the thawing soil and the undisturbed soil were extracted, respectively. Frequencies of the thawing soil and the undisturbed soil with front 8 modes are summarized in Table 7.2 and Table 7.3, respectively.

According to the results of the mode analysis, the modes of the maximum participated mass of the vertical vibration are No. 1 and No. 8. Assuming that the damping ratio remained constant, $\xi = 0.05$, according to Eq. (7.16), the damping coefficients of the thawing soil and the undisturbed soil can be calculated, as summarized in Table 7.4.

The viscous boundary conditions were conducted as the boundary conditions in this chapter. Lysmer and Kuhlemeyer proposed the concept of viscous boundary in dashpots to absorb incident waves. The dashpots were added at the boundary of the tangential and normal directions. The viscous boundary tractions in the x and y

Table 7.2 Frequency of the thawing soil with different modes

Mode	1	2	3	4	5	6	7	8
Frequency (Hz)	0.9726	1.1800	1.8950	2.1046	2.5178	2.6242	2.7840	2.9178

Table 7.3 Frequency of the undisturbed soil with different modes

Mode	1	2	3	4	5	6	7	8
Frequency (Hz)	1.0152	1.3276	2.1077	2.3853	2.7722	2.8828	3.0459	3.1781

Table 7.4 Damping coefficients of the thawing soil and the undisturbed soil

Damping coefficient	α	β
Thawing soil	0.4583	0.0041
Undisturbed soil	0.4835	0.0038

directions are calculated by the following equations:

$$f_x = -\rho c_p \frac{\partial u}{\partial t}$$
$$f_y = -\rho c_s \frac{\partial v}{\partial t} \tag{7.17}$$

where f_x and f_y are tractions applied to the boundary; ρ is the mass density; c_p and c_s are the velocities of P and S wave, respectively.

7.4 Analysis of Results

7.4.1 Vibration Acceleration of the Thawing Soil

Figure 7.4a, b, c and d illustrate the vibration acceleration of the thawing soil at 0.5 s, 1.0 s, 1.5 s and 2.0 s, respectively.

From Fig. 7.4, the direction of wave propagation (the direction of energy propagation) is mainly downward along the bottom of the tunnel. The area of larger acceleration is mainly concentrated on the lower half of the tunnel, and it is a parabolic distribution. The acceleration at the contact position of the tunnel bottom between the soil and the segment experiences the maximum, which is about 50–60 m/s^2. But with the increase of the distance from the bottom, the vibration acceleration decreases rapidly to be zero and the distance is about 10 m.

7.4.2 Amplitude-Frequency Curves of the Displacements

The amplitude-frequency curves of the displacements for the undisturbed soil and the thawing soil can be obtained after Fast Fourier Transform (FFT) of the time-history curves of the point A, B and C, as shown in Fig. 7.5 and Fig. 7.6, respectively.

Comparing the displacement spectrograms of the thawing soil and the undisturbed soil at point A, B, and C, the main frequency range of the thawing soil is larger than that of the undisturbed soil. The frequency range of the thawing soil is 0–8 Hz, while that of the undisturbed soil is 0–4 Hz. There are two peak frequencies (2 and 6 Hz) for the thawing soil at point A, B and C, while there is only one (2 Hz) for the

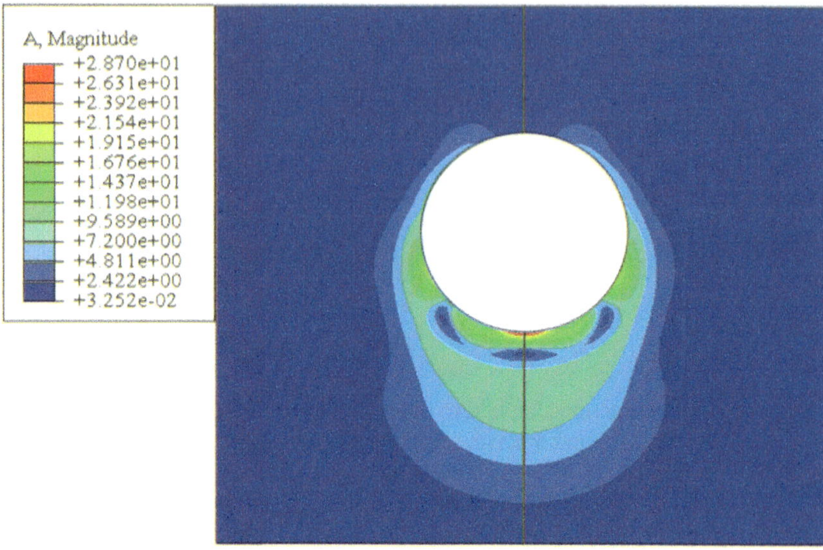

(a) 0.5s

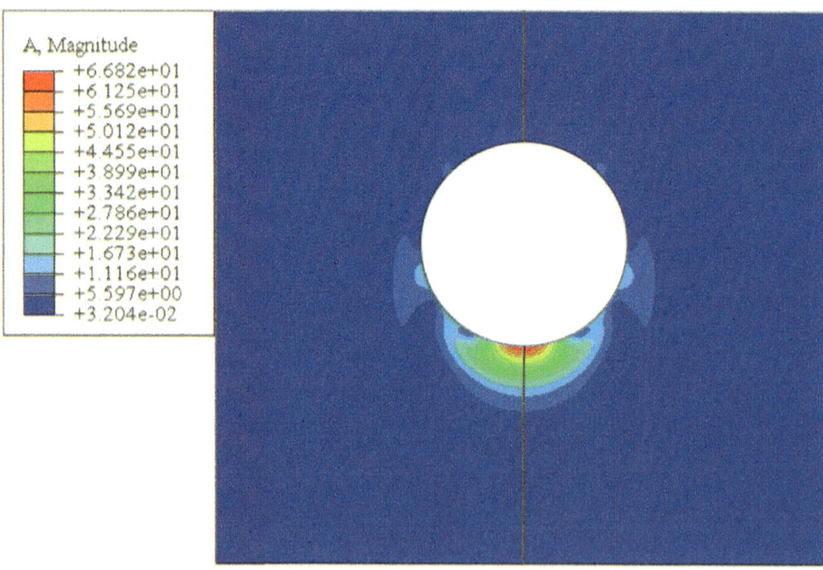

(b) 1.0 s

Fig. 7.4 Vibration acceleration of the thawing soil

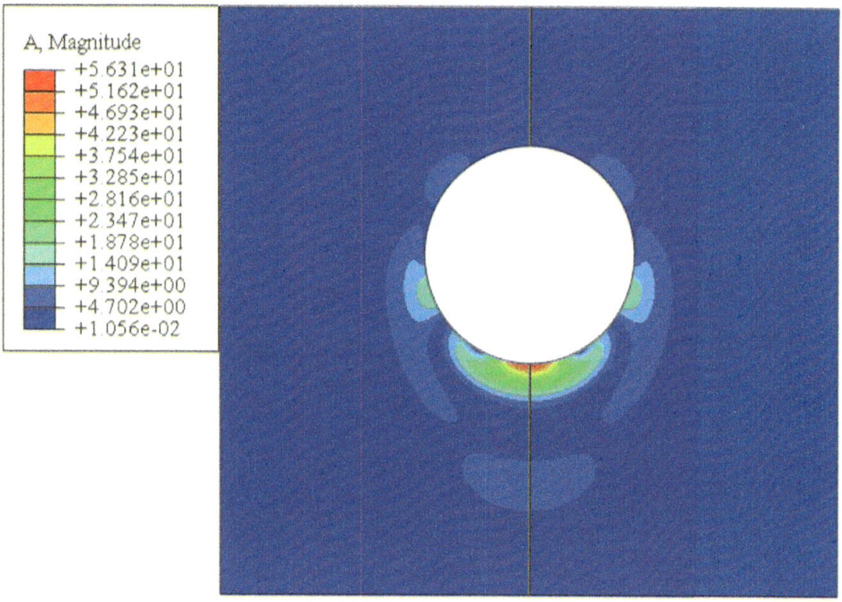

(c) 1.5s

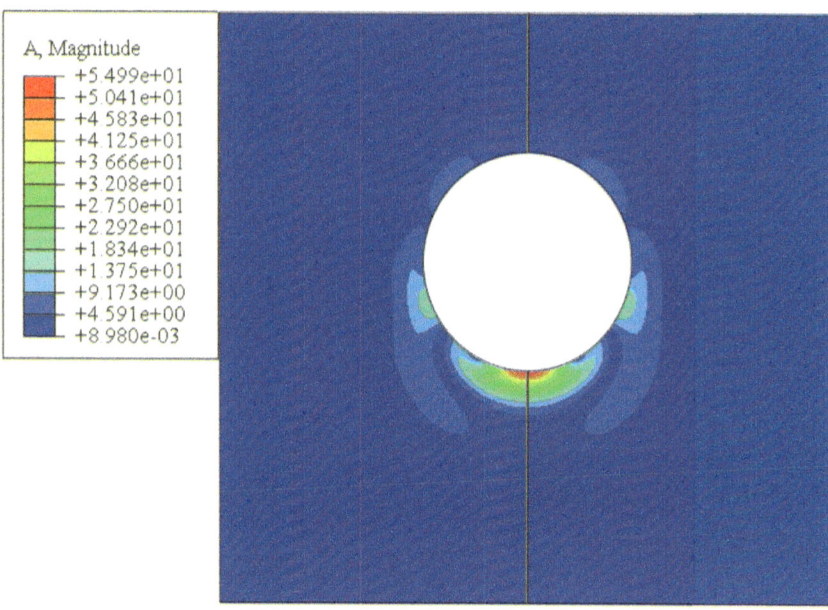

(d) 2.0s

Fig. 7.4 (continued)

Fig. 7.5 Amplitude-frequency curves of displacements at point A, B and C for undisturbed soil

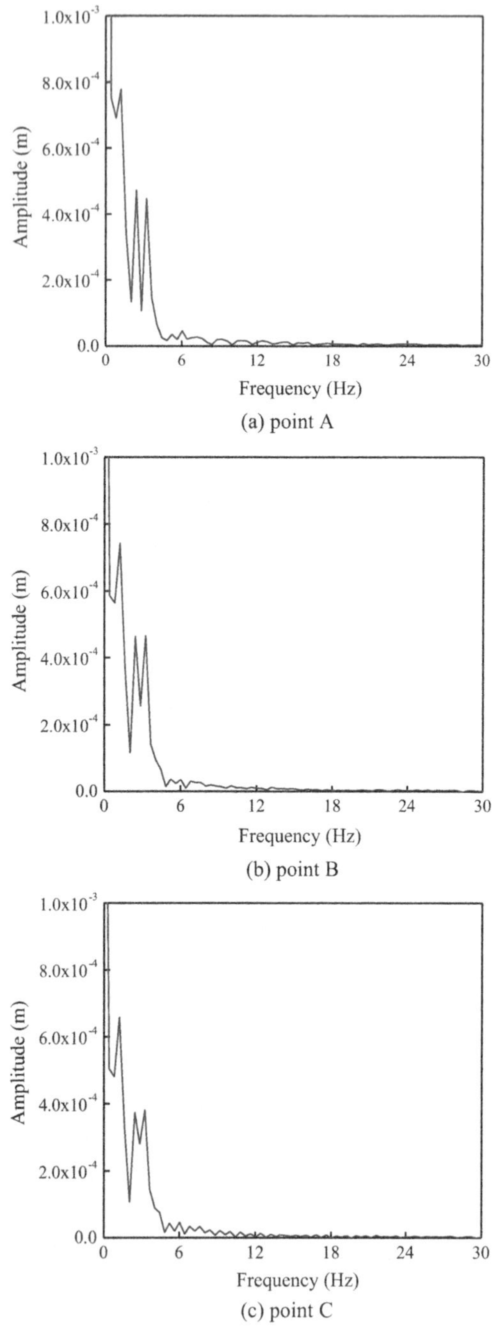

Fig. 7.6 Amplitude-frequency curves of displacements at point A, B and C for thawing soil

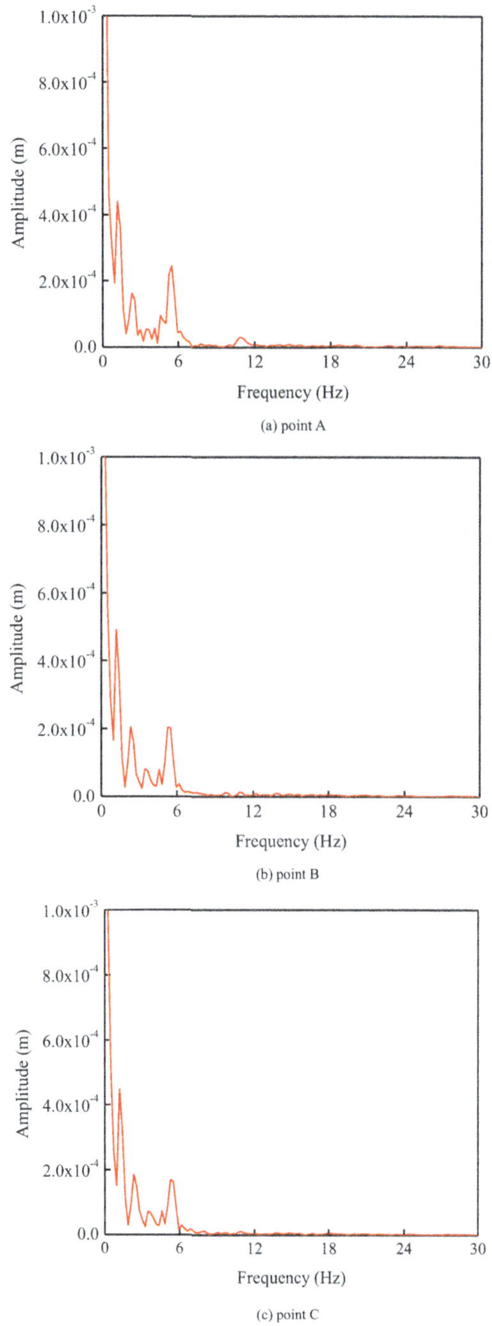

undisturbed soil. This shows that the wave in the soil decays quickly and it has been attenuated to be zero within a short distance.

7.4.3 Attenuation of the Acceleration

Figure 7.7 illustrates the attenuations of the acceleration with the distance to the vibration source in the vertical direction at 2 s. From Fig. 7.7, the thawing soil and the undisturbed soil both decay exponentially and their attenuation formulas are obtained by exponential fitting, respectively, as shown in Table 7.5.

Figure 7.7 depicts that the position of the maximum acceleration is in the vicinity of the vibration source and the acceleration of each kind of soil decays to be zero within 10 m. According to the fitting formulas, the maximum acceleration of the thawing soil is approximately 63 m/s^2, while that of the undisturbed soil is just about 30 m/s^2, 48% of the maximum acceleration of the thawing soil. Comparative analysis of the decay rates shows that the attenuation coefficients of the thawing soil and the undisturbed soil are 0.7102 and 0.4295, respectively, the attenuation coefficient of the undisturbed soil being approximately 60% of that of the thawing soil. Therefore, both the maximum vibration acceleration and the attenuation coefficient of the thawing soil are larger than those of the undisturbed soil. The main reason of the larger acceleration of the thawing soil is that the freezing and thawing decreases

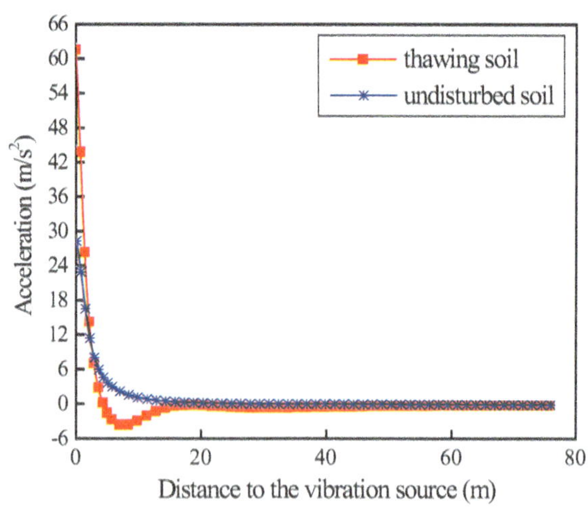

Fig. 7.7 Attenuation of acceleration with distance in the vertical direction

Table 7.5 Fitting results of the acceleration

Soil	Attenuation formulas
Thawing soil	$y = 68.2e^{-0.7102x} - 0.5315$
Undisturbed soil	$y = 29.57e^{-0.4295x} - 0.2209$

the elastic modulus and the internal soil particles are looser; the cause of the larger attenuation coefficient of the thawing soil is that the increase of the porosity enhances the absorption of vibration wave. The above analysis shows that the vibration acceleration of the thawing soil is approximately twice of that of the undisturbed soil. The soil near the segment directly affects the settlement and deformation of the tunnel. If the vibration acceleration is too large, the soil will be more prone to be damaged in the long-term cyclic loads, and the local stability of the tunnel will be affected.

7.4.4 The Maximum Ground Acceleration and Displacement

Figure 7.8 illustrates the variations of the maximum ground acceleration and the ground displacement with distance to the symmetry axis in 2 s. From Fig. 7.8a, the overall trend of the acceleration is decreasing, but there is a certain increase of the vibration acceleration of the thawing soil at 40–50 m, that is the vibration amplification appears. The reasons of the increase of the ground acceleration are that: (1) the Rayleigh wave emerges after the vibration waves propagate for a certain distance; (2) the waves with various frequencies attenuate inconsistent in the process of propagation.

The maximum ground accelerations that appears at the top of the tunnel, are approximately 1.48 m/s^2 and 0.92 m/s^2 for the thawing soil and the undisturbed soil, respectively. The maximum acceleration of the undisturbed soil is just about 62% of that of the thawing soil. But at the same time, the acceleration of the thawing soil attenuates more rapidly and it attenuates to be near 0.3 m/s^2 when the distance reaches 20 m. The maximum ground acceleration of the undisturbed soil experiences a relatively smooth change and attenuates slowly, which maintains the magnitude between 0.3 and 0.6 m/s^2. The acceleration of the undisturbed soil is larger than that of the thawing soil when the distance to the symmetry axis is more than 10 m. Meanwhile, there is no obvious amplification region on the ground surface of the undisturbed soil.

From Fig. 7.8b, the maximum ground displacements of both the thawing soil and the undisturbed soil decrease to minimum and then increase. The displacements reach the minimum at 40–50 m away from the symmetry axis and the maximums appear at the top of the tunnel, which is about 3.0 mm. The minimum ground displacements of the thawing soil and the undisturbed soil are 1.4 mm and 1.1 mm, respectively. The changes of the maximum ground displacement of the thawing soil and the undisturbed soil are basically identical in the range of 40–80 m, while the displacement of the thawing soil is larger than that of the undisturbed soil. The main reason why the curves in Fig. 7.8b increase after the minimum value is the same as that of Fig. 7.8a, namely Rayleigh wave emerges after the vibration waves propagate after the minimum value. As the energy of Rayleigh wave is mainly concentrated on the surface, the emergence of Rayleigh wave will evidently increase the ground displacement.

Fig. 7.8 Variations of the maximum ground accelerations and displacements with distance to the symmetry axis in 2 s

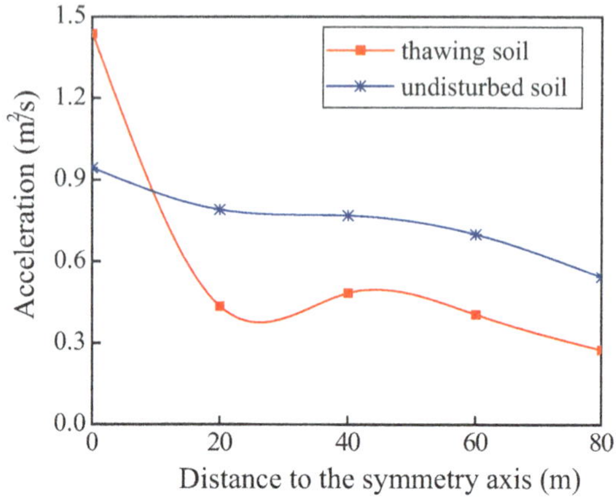

(a) maximum ground accelerations

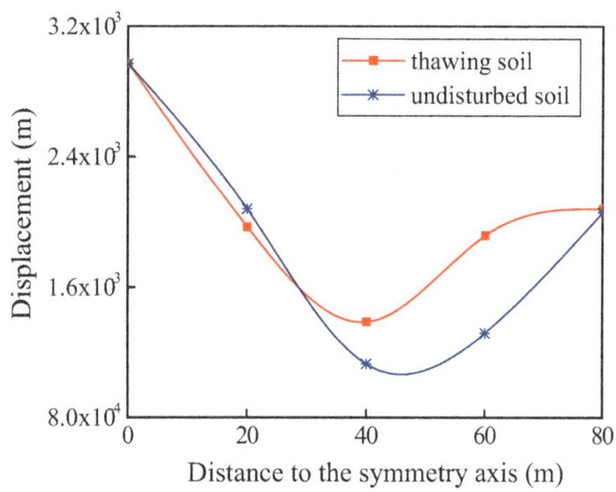

(b) maximum ground displacements

7.4.5 Time-History Curves of the Acceleration

The time-history curves of the acceleration of the thawing soil and the undisturbed soil at point A, B and C are shown in Fig. 7.8. To point A (Fig. 7.9a), there are not many differences between the thawing soil and the undisturbed soil, and the acceleration of each kind of soil fluctuates in the range of 2–12 m/s^2. to point B and C (Fig. 7.9b, c), the acceleration of the thawing soil is greater than that of the undisturbed soil. The main reasons are that the connecting forces between soil particles are weakened, and the elastic modulus and the strength of the soil reduce. Comparing the acceleration amplitudes of Fig. 7.9a–c, the acceleration amplitudes decrease gradually, which is the same as the conclusion of Fig. 7.8a.

Fig. 7.9 Variations of the acceleration with time

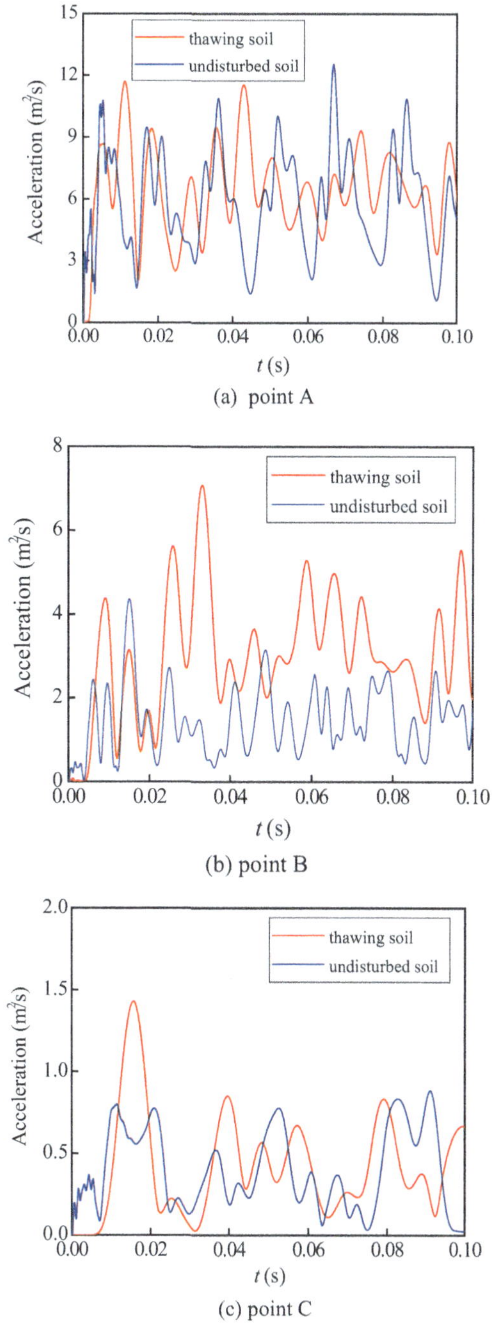

7.5 Conclusions

(1) The direction of the wave propagation is mainly downward along the bottom of the tunnel. The area of larger acceleration is mainly concentrated on the lower half of the tunnel, and it is a parabolic distribution. The maximum acceleration is at the bottom contact position between soil and segment. But with the increase of the distance from the bottom, the vibration acceleration decreases rapidly to be zero.

(2) The main frequency range of the thawing soil is larger than that of the undisturbed soil at the corresponding points. The frequency range of the thawing soil is 0–8 Hz, and the frequency range of the undisturbed soil is 0–4 Hz. There are two peak frequencies (2 and 6 Hz) of the thawing soil, while there is only one (2 Hz) of the undisturbed soil.

(3) The position of the maximum acceleration is in the vicinity of the vibration source and the acceleration of each kind of soil decays to be zero within 10 m. The maximum acceleration of the thawing soil is approximately 63 m/s^2, while it is just about 30 m/s^2 (48% of that of the thawing soil) of the undisturbed soil. The attenuation coefficients of the thawing soil and the undisturbed soil are 0.7102 and 0.4295, respectively. The attenuation coefficient of the undisturbed soil is approximately 60% of that of the thawing soil. The maximum vibration acceleration and the attenuation coefficient of the thawing soil are both larger than those of the undisturbed soil.

(4) There is a certain increase of the vibration acceleration of the thawing soil at 40–50 m, while there is no obvious amplification region of the ground surface of the undisturbed soil. The maximum accelerations at the top of the tunnel are approximately 1.48 m/s^2, 0.92 m/s^2, respectively. The maximum acceleration of the undisturbed soil is just about 62% of that of the thawing soil. But at the same time, the acceleration of the thawing soil attenuates more rapidly. It attenuates to be near 0.3 m/s^2 when the distance reaches 20 m. The maximum ground acceleration of the undisturbed soil experiences a relatively smooth change and attenuates slowly, which maintains the magnitude of 0.3–0.6 m/s^2. The acceleration of the undisturbed soil is larger than that of the thawing soil when the distance to the symmetry axis is more than 10 m.

(5) The maximum ground displacements of both the thawing soil and the undisturbed soil decrease to minimum and then increase. Each kind of soil reaches the minimum in the range of 40–50 m, and the maximum displacement appears at the top of the tunnel, which is about 3.0 mm. The minimum ground displacements of the thawing soil and the undisturbed soil are 1.4 mm and 1.1 mm, respectively. The changes of the maximum ground displacement of the thawing soil and the undisturbed soil are basically identical in the range of 40–80 m, while the displacement of the thawing soil is larger than that of the undisturbed soil.

(6) There are not many differences between the thawing soil and the undisturbed soil in the near field. But in the far field, as a result of the weaker connecting forces and smaller elastic modulus, the acceleration of the thawing soil is greater than that of the undisturbed soil.

References

Alabi, B. (1992). A parametric study on some aspects of ground-borne vibrations due to rail traffic. *Journal of Sound and Vibration, 153*(1), 77–87.

Brams, B. B., & Yao, L. Y. C. (1964). Shear strength of a soil after freezing and thawing. *Journal of Soil Mechanics and Foundation Division, 90,* 1–25.

Clouteau, D., Arnst, M., Al-Hussaini, T. M., et al. (2005). Freefield vibrations due to dynamic loading on a tunnel embedded in a stratified medium. *Journal of Sound and Vibration, 283*(1), 173–199.

Dawn, T. M., & Stanworth, C. G. (1979). Ground vibrations from passing trains. *Journal of Sound and Vibration, 66*(3), 355–362.

Dieterman, H. A., & Metrikine, V. (1997). Steady-state displacements of a beam on an elastic half-space due to a uniformly moving constant load. *European Journal of Mechanics Series A Solids, 16,* 295–306.

Eason, G. (1965). The stresses produced in a semi-infinite solid by a moving surface force. *International Journal of Engineering Science, 2*(6), 581–609.

Ewing, W. M., Jardetzky, W. S., & Press, F. (1958). Elastic waves in layered media. *GFF, 80*(1), 128–129.

Finn, W. D., & Yong, R. N. Y. (1978). Seismic response of frozen ground. *Journal of the Geotechnical Engineering Division, 104*(10), 1225–1241.

Forrest, J. A., & Hunt, H. E. M. (2006). A three-dimensional tunnel model for calculation of train-induced ground vibration. *Journal of Sound and Vibration, 294,* 678–705.

Graham, J., & Au, V. C. S. (1985). Effects of freeze-thaw and softening on a natural clay at low stresses. *Canadian Geotechnical Journal, 22*(1), 69–78.

Guan, F., & Moore, I. D. (1994). Three-dimensional dynamic response of twin cavities due to travelling loads. *Journal of Engineering Mechanics, 120*(3), 637–651.

He, P. P., & Cui, Z. D. (2015). Dynamic response of a thawing soil around the tunnel under the vibration load of subway. *Environmental Earth Sciences, 73*, 2473–2482.

Hussein, M. F. M., & Hunt, H. E. M. (2007). A numerical model for calculating vibration from a railway tunnel embedded in a full-space. *Journal of Sound and Vibration, 305,* 401–431.

Israil, A. S. M., & Ahmad, S. (1989). Dynamic vertical compliance of strip foundations in layered soils. *Earthquake Engineering and Structural Dynamics, 18*(7), 933–950.

Jenkins, H. H., Stephenson, J. E., Clayton, G. A., et al. (1974). The effect of track and vehicle parameters on wheel/rail vertical dynamic forces. *Railway Engineering Journal, 3*(1), 2–16.

Kaynia, A. M., Madshus, C., & Zackrisson, P. (2000). Ground vibration from high-speed trains: prediction and countermeasures. *Journal of Geotechnical and Geoenvironmental Engineering, 126*(6), 531–537.

Konrad, J. M. (1989). Physical processes during freeze-thaw cycles in clayey silts. *Cold Regions Science and Technology, 16*(3), 291–303.

Lieb, M., & Sudret, B. (1998). A fast algorithm for soil dynamics calculations by wavelet decomposition. *Archive of Applied Mechanics, 68*(3–4), 147–157.

Lu, J. F., Jeng, D. S., & Lee, T. L. (2007). Dynamic response of a piecewise circular tunnel embedded in a poroelastic medium. *Soil Dynamics and Earthquake Engineering, 27,* 875–891.

Nejati, H. R., Ahmadi, M., & Hashemolhosseini, H. (2012). Numerical analysis of ground surface vibration induced by underground train movement. *Tunnelling and Underground Space Technology, 29,* 1–9.
Pan, C. S., & Pande, G. N. (1984). Preliminary deterministic finite element study on a tunnel driven in loess subjected to train loading. *Journal of Civil Engineering, 17*(4), 18–28. (in Chinese with English abstract).
Rieckh, G., Kreuzer, W., Waubke, H., et al. (2012). A 2.5 D-Fourier-BEM model for vibrations in a tunnel running through layered anisotropic soil. *Engineering Analysis with Boundary Elements, 36*(6), 960–967.
Sheng, X., Jones, C. J. C., & Thompson, D. J. (2002). Ground vibration generated by a harmonic load moving in a circular tunnel in a layered ground. In *Proceedings of the 10th International Meeting on Low Frequency Noise and Vibration and Its Control* (pp. 83–96).
Sheng, X., Jones, C. J. C., & Thompson, D. J. (2006). Prediction of ground vibration from trains using the wavenumber finite and boundary element methods. *Journal of Sound and Vibration, 293*(3), 575–586.
Singh, S., & Donovan, N. C. (1977). Seismic response of frozen-thawed soil systems. In *Proceedings of Sixth World Conference on Earthquake Engineering, Sarita Prakashan, Meerut, India* (pp. 2262–2226).
Tang, Y. Q., Yang, Q., & Yu, H. (2014). Changes of the pore distribution of silty clay under the subway train loads. *Environmental Earth Sciences.*
Ting, J. M., Martin, R. T., & Ladd, C. C. (1983). Mechanisms of strength for frozen sand. *Journal of the Geotechnical Engineering Division, 109*(10), 1286–1302.
Viklander, P., & Eigenbrod, D. (2000). Stone movements and permeability changes in till caused by freezing and thawing. *Cold Regions Science and Technology, 31*(2), 151–162.
Vinson, T. S., Chaichanavong, T., & Czajkowski, R. L. (1978). Behavior of frozen clays under cyclic axial loading. *Journal of the Geotechnical Engineering Division, 104,* 779–780.
Vinson, T. S., & Li, J. C. (1980). Dynamic properties of frozen sand under simulated earthquake loading conditions. In *Proceedings of the Seventh World Conference on Earthquake Engineering. Istanbul* (pp. 225–240).
Wolf, J. P. (1985). *Dynamic soil-structure interaction.* Englewood Cliffs: Prentice-Hall.
Wolf, J. P. (1988). *Soil-structure interaction analysis in time domain.* A Division of Simon & Schuster.
Yan, C. L., Tang, Y. Q., & Liu, Y. T. (2013). Study on fractal dimensions of the silty soil around the tunnel under the subway loading in Shanghai. *Environmental Earth Sciences, 69,* 1529–1535.
Yang, P., & Zhang, T. (2002). Research on the physical and mechanical properties of the artificial thawing soil. *Glaciol, 24*(5), 665–667. (in Chinese).
Yang, Y. B., Hung, H. H., & Chang, D. W. (2003). Train-induced wave propagation in layered soils using finite/infinite element simulation. *Soil Dynamics and Earthquake Engineering, 23*(4), 263–278.

Chapter 8
Conclusions and Prospects

8.1 Conclusions

In this monograph, the dynamic characteristics of soil around subway shield tunnel by artificial freezing method were studied, which can provide theoretical basis and guidance for the effective control of long-term subway settlement and maintaining the safety of subway operation. The results are as follows.

(1) The freezing parameters of the by-pass are designed, mainly including the thickness and the average temperature of the freezing wall, the freezing hole, the brine temperature and the flow rate. The force-method and finite element software are used to design the thickness of the freezing wall. The former result is 3.5 m while the latter demonstrates that the freezing wall with thickness of 1.8 m can meet the requirements, which indicates that the force-method is conservative. The total number of the freezing holes is 74. The type of the freezing pipe is $\Phi 89 \times 8$. The design value of the average temperature of the freezing wall is -10 °C. The lowest brine temperature must not be higher than -28°C. The brine flow of the single hole shouldn't be smaller than 3 m^3/h. During the freezing period, the descent speed of the soil temperature and brine temperature is large at the beginning. Then it becomes small gradually and finally tends to be stable. The temperature of the soil inside the freezing wall drops faster and the final temperature is lower, so is the temperature of the soil on the side the freezing station. The soil at the center of the by-pass is subjected to the largest influence of the frost heave, so is the growth rate of the frost heave. The farther away from the center, the smaller the influence of frost heave, so does the growth rate of frost heave. According to the measured data, the calculated thickness of the freezing wall is 26–80% larger than the designed thickness, while the average temperature of the freezing wall is only 11–21% larger than the designed thickness. This result indicates that compared with the thickness of the freezing wall, the average temperature is more difficult to meet the design requirements.

Z.-D. Cui et al., *Dynamics of Freezing-Thawing Soil around Subway Shield Tunnels*,
https://doi.org/10.1007/978-981-15-4342-5_8

(2) H-D model and M-D model both belong to the modified hyperbolic Iwan model. M-D model is relatively accurate and agrees better with the test data for freezing-thawing and undisturbed samples comparing with H-D model, even though they both meet the requirements and the results are very close. Pore water becomes ice with volume expansion during freezing, resulting in the increase of porosity of clay samples. It is equivalent to produce a certain disturbance to the undisturbed clay. Cyclic loads aggravate the destruction of clay structures after freezing and thawing. So the freezing-thawing clay can produce up to 26.7% larger strain and 53% higher excess pore pressure than that of undisturbed clay under the same tests conditions, respectively. The rates of axial strain and excess pore pressure decrease gradually with the number of vibration increasing. The frequency and the CSR have significant influence on the dynamic behaviors of the saturated soft clay. The lower the frequency and the larger the amplitude of deviator stress are, the more energy transferred to soil is; the larger the axial strain and the higher excess pore pressure are developed, the earlier the inflection points of deformation appear.

(3) The damping ratio of soil represents the capability of samples to absorb energy under cyclic loading. The larger the damping ratio is; the more energy is absorbed by samples. The energy transformed into soil partly leads the plastic deformation of samples and partly increases the excess pore pressure under the cyclic loading. The energy can be represented by the areas enclosed by the load and unload hysteretic curves, which are well fitted by the geometric hyperbolic functions. The frequency and the CSR have significant effects on the dynamic behaviors of the saturated soft clay. The lower the frequency and the larger the CSR are, the bigger the dynamic modulus degradation is. The modulus decreases significantly when the strain is small. The dynamic modulus increased after freezing-thawing under the same cyclic loading condition. The temperature of −10 °C can be considered as an inflected point of freezing temperature that determines the dynamic characteristics of freezing-thawing clay. When the temperature drops to −10 °C in the freezing process, the disturbance from ice extrusion to clay structures is the most serious after thawing. The modulus of the freezing-thawing sample with −30 °C is 12.2% larger than that of the undisturbed sample under the same conditions. With the increasing of number of vibration, the damping ratio curves show a downward trend. The decline of damping ratio before and after triaxial tests increases with the freezing temperature dropping. The lower the temperature is, the greater the pore volumes are expanded and more room is obtained for deformation of soil samples after thawing. For the freezing-thawing sample with temperature of −30 °C, the damping ratio drops about 50% after cyclic triaxial tests. With the increasing of number of vibration, the increment of deformation and the excess pore water pressure within samples under the cyclic loading become smaller and smaller. The lower the frequency and the larger the amplitude of deviator stress are, the larger the axial strain and the higher excess pore pressure develop. But the rates decrease gradually with the number of vibration increasing. The excess pore water pressure and axial strain increase after freezing-thawing. The lower

the freezing temperature is, the higher the excess pore water pressure develops. The axial strains of samples with CSRs of 0.25 and 0.125 increase 16% and 5% after freezing-thawing, respectively.

(4) The corrected Otsu method was used to obtain the threshold for the SEM images of silty clay. The freezing-thawing clay exhibits larger aggregates and larger pore structures than those of the undisturbed clay. After cyclic loading, the microstructures of the particle aggregates are compacted. The freezing-thawing process has little effect on the pore shape coefficient of the silty clay, but the mean anisotropy value of the microscopic pore structures increases about 12% after freezing-thawing. The pore orientation of samples without loading are arranged horizontally, while the pores of samples under cyclic loadings are arranged vertically. The porosity increases with the pore diameter increasing in the range of 1.0–8.0 μm. The porosity of samples without loadings increases about 11.24% after freezing-thawing. The lower the freezing temperature is, the larger the porosity of the samples is. There is no obvious tendency of the porosity distribution with the axial strain, while the porosity decreases with the excess pore pressure increasing. The microscopic pore structures of the silty clay exhibit fractal characteristics with the fractal dimension between [1.15, 1.27]. The fractal dimension is reduced by the disturbance from the freezing-thawing effect, coupled with the cyclic loading.

(5) The mercury intrusion pressure is inversely related to the pore radius of the soil. There exists a bottleneck effect in the mercury intrusion process. The mercury volume increases slowly at both the beginning and the ending of the MIP tests and increases rapidly at the pore radii of 0.1–0.6 μm. In the extrusion process, some of the mercury is trapped in pores with narrow necks which leads to hysteresis of the curves. The log-differential pore volume curves of the silty clay show a unimodal mode, where the mode with pore radii range from 0.05 to 0.6 μm represents intra-aggregate pores. The water in the saturated soil turns into ice with the volume expanding when the temperature is lowered below its freezing point. After thawing, the ice in the soil changes to be water, which leaves enlargement to the microscope pore structures of the silty clay. For the samples without cyclic loadings, the final mercury intrusion volume and the most probable pore size increase about 6.0% and 30.64% after freezing-thawing, respectively. Under the same cyclic loading, the mercury intrusion volume increases with the freezing temperature decreasing and the most probable pore sizes are close to each other. The cyclic loadings have great effects on the PSD of the silty clay. Under the cyclic loading pressure, the pore volumes are contracted. The lower the frequency and the larger the CSR of cyclic loadings are, the larger the deformations of the samples are and the smaller the most probable pore size becomes. The pore volume of freezing-thawing samples with the frequency of 2.5 Hz is 8.74% larger than that with 0.5 Hz, while it is 5.03% of undisturbed samples. The most probably pore size of the freezing-thawing samples with the CSR of 0.125 is about 81.2% larger than that of the CSR of 0.375. The process of mercury intrusion in a porous medium is similar to air injection in a saturated soil as described along the drying water retention curve. The mercury intrusion

pressure and the normalized volume are equivalent to the soil suction and the air volume within the soil, respectively. The MIP derived WRC are suitable for the prediction of the soil-water characteristic curves with the van Genuchten model. In addition, the microscope pore structures of the silty clay exhibit fractal characteristics with the thermal fractal dimension model in MIP tests.

(6) The direction of the wave propagation is mainly downward along the bottom of the tunnel. The area of larger acceleration is mainly concentrated on the lower half of the tunnel, and it is a parabolic distribution. The maximum acceleration is at the bottom contact position between soil and segment. But with the increase of the distance from the bottom, the vibration acceleration decreases rapidly to be zero. The main frequency range of the thawing soil is larger than that of the undisturbed soil at the corresponding points. The frequency range of the thawing soil is 0–8 Hz, and the frequency range of the undisturbed soil is 0–4 Hz. There are two peak frequencies (2 and 6 Hz) of the thawing soil, while there is only one (2 Hz) of the undisturbed soil. The position of the maximum acceleration is in the vicinity of the vibration source and the acceleration of each kind of soil decays to be zero within 10 m. The maximum acceleration of the thawing soil is approximately 63 m/s^2, while it is just about 30 m/s^2 (48% of that of the thawing soil) of the undisturbed soil. The attenuation coefficients of the thawing soil and the undisturbed soil are 0.7102 and 0.4295, respectively. The attenuation coefficient of the undisturbed soil is approximately 60% of that of the thawing soil. The maximum vibration acceleration and the attenuation coefficient of the thawing soil are both larger than those of the undisturbed soil. There is a certain increase of the vibration acceleration of the thawing soil at 40–50 m, while there is no obvious amplification region of the ground surface of the undisturbed soil. The maximum accelerations at the top of the tunnel are approximately 1.48 m/s^2, 0.92 m/s^2, respectively. The maximum acceleration of the undisturbed soil is just about 62% of that of the thawing soil. But at the same time, the acceleration of the thawing soil attenuates more rapidly. It attenuates to be near 0.3 m/s^2 when the distance reaches 20 m. The maximum ground acceleration of the undisturbed soil experiences a relatively smooth change and attenuates slowly, which maintains the magnitude of 0.3–0.6 m/s^2. The acceleration of the undisturbed soil is larger than that of the thawing soil when the distance to the symmetry axis is more than 10 m. The maximum ground displacements of both the thawing soil and the undisturbed soil decrease to minimum and then increase. Each kind of soil reaches the minimum in the range of 40–50 m, and the maximum displacement appears at the top of the tunnel, which is about 3.0 mm. The minimum ground displacements of the thawing soil and the undisturbed soil are 1.4 mm and 1.1 mm, respectively. The changes of the maximum ground displacement of the thawing soil and the undisturbed soil are basically identical in the range of 40–80 m, while the displacement of the thawing soil is larger than that of the undisturbed soil. There are not many differences between the thawing soil and the undisturbed soil in the near field.

But in the far field, as a result of the weaker connecting forces and smaller elastic modulus, the acceleration of the thawing soil is greater than that of the undisturbed soil.

8.2 Prospects for Further Study

This monograph studied the dynamic characteristics of soil around subway shield tunnel by artificial freezing method. But there is still some work to be further studied.

(1) Sensitivity analysis should be carried out to study the influence of loading frequency, dynamic stress amplitude, loading times and freezing temperature on dynamic characteristics of soft clay.
(2) It is necessary to study the dynamic constitutive model of soil, find the suitable dynamic constitutive model of clay in soft soil area, and carry out the secondary development of finite element.
(3) Through high-precision 3D printing technology, the microstructure of soil can be simulated, and the interaction between particles can be studied to reveal the mechanism of microstructure change of soft clay, and the relationship between microstructure change and macro variables can be studied.

Appendix
Major Published Works of the Book Author

Cui, Z. D., is noted the corresponding author.

Z.-D. Cui et al., *Dynamics of Freezing-Thawing Soil around Subway Shield Tunnels*,
https://doi.org/10.1007/978-981-15-4342-5

Bibliography

Monograph, Textbook and Proceedings

Cui, Z. D. (2018). *Land subsidence induced by the engineering-environmental effects*. Springer Press. ISBN 978-981-8039-5.

Cui, Z. D., & Zhang, Z. L. (2017). *Design of underground structures* (in Chinese). China Construction Industry Press. ISBN 978-7-112-20816-6.

Cui, Z. D., Zhang, Z. L., Yuan, L., et al. (2019). *Design of underground structures*. Springer Press. ISBN 978-981-13-7731-0.

Kallel, A., Erguler, Z. A., Cui, Z. D., et al. (2019). Recent advances in geo-environmental engineering, geomechanics and geotechnics, and geohazards. In *Proceedings of the 1st Springer Conference of the Arabian Journal of Geosciences (CAJG-1)*. Springer Press: Tunisia 2018. ISBN 978-3-030-01664-7.

English Journal Papers

Cui, Z. D. (2011). Effect of water–silt composite blasting on the stability of rocks surrounding a tunnel. *Bulletin of Engineering Geology and the Environment (SCI), 70*(4), 657–664.

Cui, Z. D. (2012a). Bearing capacity of single pile and in-flight T-Bar Penetration for centrifuge modeling of land subsidence caused by the interaction of high-rise buildings. *Bulletin of Engineering Geology and the Environment, 71*(3), 579–586.

Cui, Z. D. (2012b). Land subsidence disaster caused by natural factors and human activities. *Disaster Advances (SCI), 5*(2), 3–4.

Cui, Z. D., & Tang, Y. Q. (2010). Land subsidence and pore structure of soils caused by the high-rise building group through centrifuge model test. *Engineering Geology (SCI), 113*(1–4), 44–52.

Cui, Z. D., & Wang, H. M. (2010). Land subsidence at different points among a group of high-rise buildings. *Disaster Advances (SCI), 3*(4), 63–66.

Cui, Z. D., & Tang, Y. Q. (2011). Microstructures of different soil layers caused by the high-rise building group in Shanghai. *Environmental Earth Sciences (SCI), 63*(1), 109–119.

Cui, Z. D., & Ren, S. X. (2012). Long-term deformation characteristics of the soil around the subway tunnel induced by the vibration loading. *Disaster Advances (SCI), 5*(4), 1791–1797.

Z.-D. Cui et al., *Dynamics of Freezing-Thawing Soil around Subway Shield Tunnels*, https://doi.org/10.1007/978-981-15-4342-5

Cui, Z. D., & Jia, Y. J. (2012). Study on the mechanisms of the soil consolidation and land subsidence caused by the high-rise building group in the soft soil area. *Disaster Advances (SCI), 5*(4), 604–608.

Cui, Z. D., & Jia, Y. J. (2013). Analysis of electron microscope images of soil pore structure for the study of land subsidence in centrifuge model tests of high-rise building groups. *Engineering Geology (SCI), 164*, 107–116.

Cui, Z. D., & Ren, S. X. (2014). Prediction of long-term settlements of subway tunnel in the soft soil area. *Natural Hazards (SCI), 74*(2), 1007–1020.

Cui, Z. D., & Tan, J. (2015). Study on settlements of the running tunnels and subway stations based on the in-situ monitoring data. *Natural Hazards (SCI), 75*(1), 465–472.

Cui, Z.D., & Yuan, Q. (2015). Investigation of the subsidence induced by the Maglev train. *Natural Hazards (SCI), 75*(2), 1767–1778.

Cui, Z. D., & Zhang, Z. L. (2015). Comparison of dynamic characteristics of the silty clay before and after freezing and thawing under the subway vibration loading. *Cold Regions Science and Technology (SCI), 119*, 29–36.

Cui, Z. D., & Jia, Y. J. (2018). Physical model test of layered soil subsidence considering dual effects of building load and groundwater withdrawal. *Arabian Journal for Science and Engineering (SCI), 48*, 1721–1734.

Cui, Z. D., Tang, Y. Q., Guo, C. Q., Yuan, L., & Yan, C. L. (2008). Flow-induced vibration and stability of an element model for parallel-plate fuel assemblies. *Nuclear Engineering and Design (SCI), 238*(7), 1629–1635.

Cui, Z. D., Tang, Y. Q., & Yan, X. X. (2010a) Centrifuge modeling of land subsidence caused by the high-rise building group in the soft soil area. *Environmental Earth Sciences (SCI), 59*(8), 1819–1826.

Cui, Z. D., Yuan, L., & Yan, C. L. (2010b). Water-silt composite blasting for tunneling. *International Journal of Rock Mechanics and Mining Sciences (SCI), 47*(6), 1034–1037.

Cui, Z. D, Tang, Y. Q., Yan, X. X., Wang, H. M., Yan, C. L., & Wang, J. X. (2010c). Evaluation of the geology-environmental capacity of buildings based on the ANFIS model of the floor area ratio. *Bulletin of Engineering Geology and the Environment (SCI), 69*(1), 111–118.

Cui, Z. D., He, P. P., & Yang, W. H. (2014). Mechanical properties of a silty clay subjected to freezing-thawing. *Cold Regions Science and Technology (SCI), 98*, 26–34.

Cui, Z. D., Yang, J. Q., & Yuan, L. (2015). Land subsidence caused by the interaction of high-rise buildings in soft soil areas. *Natural Hazards (SCI), 79*(2), 1199–1217.

Cui, Z. D., Jia, Y. J., & Yuan, L. (2016a). Distribution law of soil deformation caused by decompression of confined water. *Environmental Earth Sciences (SCI), 75*(18), 1281.

Cui, Z. D., Li, Z., & Jia, Y. J. (2016b). Model test study on the subsidence of high-rise building group due to the variation of groundwater level. *Natural Hazards (SCI), 84*(1), 35–53.

Cui, Z. D., Zhao, L. Z., & Yuan, L. (2016c). Microstructures of consolidated Kaolin clay at different depths in centrifuge model tests. *Carbonates and Evaporites (SCI), 31*(1), 47–60.

Cui, Z. D., Yuan, Q., & Yang, J. Q. (2018). Laboratory model tests about the sand embankment supported by piles with a cap beam. *Geomechanics and Geoengineering an International Journal, 13*(1), 64–76.

He, P. P., & Cui, Z. D. (2015). Dynamic response of a thawing soil around the tunnel under the vibration load of subway. *Environmental Earth Sciences (SCI), 73*(5), 2473–2482.

Li, Z., & Cui, Z. D. (2017). Axisymmetric consolidation of saturated multi-layered soils with anisotropic permeability due to well pumping. *Computers and Geotechnics (SCI), 92*, 229–239.

Song, L., Cui, Z. D., & Zhang, H. Q. (2013). Analysis and treatment of the fault activation below the dynamic foundation in the goaf area. *Disaster Advances (SCI), 6*(S1), 337–342.

Tan, J., & Cui, Z. D., & Yuan, Y. (2016). Study on the long-term settlement of subway tunnel in soft soil area. *Marine Georesources & Geotechnology (SCI), 34*(5), 486–492.

Tang, Y. Q., Cui, Z. D., Wang, J. X., Lu, C., & Yan, X. X. (2008). Model test study of land subsidence caused by high-rise building group. *Bulletin of Engineering Geology and the Environment (SCI), 67*(2), 173–179.

Tang, Y. Q., Cui, Z. D., Wang, J. X., Yan, L. P., & Yan, X. X. (2008a). Application of grey theory-based model to prediction of land subsidence due to engineering environment in Shanghai. *Environmental Geology (SCI), 55*(3), 583–593.

Tang, Y. Q., Cui, Z. D., Zhang, X., & Zhao, S. K. (2008b). Dynamic response and pore-water pressure model of saturated soft clay around a tunnel induced by the subway vibration load. *Engineering Geology (SCI), 98*(3–4), 126–132.

Xu, X. B., & Cui, Z. D. (2020). Investigation of a fractional derivative creep model of clay and its numerical implementation. *Computers and Geotechnics (SCI), 119*, 103387.

Yuan, L., & Cui, Z. D. (2013). Reliability analysis for the consecutive-k-out-of-n: F system with repairmen taking multiple vacations. *Applied Mathematical Modelling (SCI), 37*(7), 4685–4697.

Yuan, Q., & Cui, Z. D. (2016). Two-dimensional numerical analysis of the subgrade improved by stone columns in the soft soil area. *Marine Georesources & Geotechnology (SCI), 34*(1), 79–86.

Yuan, L., Cui, Z. D., & Tan, J. (2017). Numerical simulation of longitudinal settlement of shield tunnel in the Coastal City, Shanghai. *Marine Georesources & Geotechnology (SCI), 35*(3), 365–370.

Zhang, C. L., & Cui, Z. D. (2017a). Numerical simulation of dynamic response around shield tunnel in the soft soil area. *Marine Georesources & Geotechnology (SCI), 35*(7), 1018–1027.

Zhang, Z. L., & Cui, Z. D. (2017b). Analysis of microscopic pore structures of the silty clay before and after freezing–thawing under the subway vibration loading. *Environmental Earth Sciences (SCI), 76*, 528.

Zhang, Z. L., & Cui, Z. D. (2018a). Effect of freezing-thawing on dynamic characteristics of the silty clay under K0-consolidated condition. *Cold Regions Science and Technology (SCI), 146*, 32–42.

Zhang, Z. L., & Cui, Z. D. (2018b). Effects of freezing-thawing and cyclic loading on pore size distribution of silty clay by mercury intrusion porosimetry. *Cold Regions Science and Technology, 145*, 185–196.

English Conference Papers

Cui, Z. D. (2019). Land subsidence induced by the engineering-environmental effect in Shanghai, China. In *Proceedings of the 1st Springer Conference of the Arabian Journal of Geosciences (CAJG-1): Recent Advances in Geo-Environmental Engineering, Geomechanics and Geotechnics, and Geohazards* (pp. 11–14).

Cui, Z. D., & Zhang, Z. L. (2017). Comparison of the dynamic characteristics of the silty clay before and after freezing-thawing under cyclic loadings. In *Seoul: 19th ICSMGE* (pp. 335–339).

Cui, Z. D., & Hua, S. S. (2018). Long-term settlement of subway tunnel and prediction of settlement trough in Coastal City Shanghai. In *Proceedings of GeoShanghai 2018 International Conference: Multi-physics Processes in Soil Mechanics and Advances in Geotechnical Testing* (pp. 458–467).

Cui, Z. D., & Fan, S. C. (2018). Physical modeling of arching effect in the piled embankment. In *Proceedings of GeoShanghai 2018 International Conference: Fundamentals of Soil Behaviours* (pp. 379–387).

Cui, Z. D., Tang, Y. Q., & Zhang, X. (2008). Deformation and pore pressure model of saturated soft clay around a subway tunnel. In *Shanghai: The Six International Symposium Geotechnical Aspects of Underground Construction in Soft Ground (EI)* (pp. 769–774).

Cui, Z. D., Yuan, L., Yan, L. P., & Guo, C. Q. (2009). Stability analysis and measurement of flow-induced vibration of a parallel-plate structure. In *Chengdu: International Conference on Earthquake Engineering-the 1st Anniversary of Wenchuan Earthquake (ISTP)* (pp. 288–293).

Cui, Z. D., Zhang, C. L., & Hou, C. Y. (2017). *Vibrations induced by subway moving load with excitation of the track vertical profile irregularity*. In *Wuhan: 15th IACMAG* (pp. 1–18).

Tang, Y. Q., Cui, Z. D., & Zhang, X. (2008). Dynamic response of saturated silty clay around a tunnel under subway vibration loading in Shanghai. In *Shanghai: The Six International Symposium Geotechnical Aspects of Underground Construction in Soft Ground (EI)* (pp. 843–848).

Zhang, Z. L., & Cui, Z. D. (2018). Dynamic response of soil around the tunnel under subway vibration loading. In *Proceedings of GeoShanghai 2018 International Conference: Advances in Soil Dynamics and Foundation Engineering* (pp. 53–61).